Studies in Fuzziness and Soft Computing

Volume 333

Series editor

Janusz Kacprzyk, Polish Academy of Sciences, Warsaw, Poland
e-mail: kacprzyk@ibspan.waw.pl

About this Series

The series "Studies in Fuzziness and Soft Computing" contains publications on various topics in the area of soft computing, which include fuzzy sets, rough sets, neural networks, evolutionary computation, probabilistic and evidential reasoning, multi-valued logic, and related fields. The publications within "Studies in Fuzziness and Soft Computing" are primarily monographs and edited volumes. They cover significant recent developments in the field, both of a foundational and applicable character. An important feature of the series is its short publication time and world-wide distribution. This permits a rapid and broad dissemination of research results.

More information about this series http://www.springer.com/series/2941

Iwona Skalna · Bogdan Rębiasz
Bartłomiej Gaweł · Beata Basiura
Jerzy Duda · Janusz Opiła
Tomasz Pełech-Pilichowski

Advances in Fuzzy Decision Making

Theory and Practice

Iwona Skalna
AGH University of Science and Technology
Kraków
Poland

Jerzy Duda
AGH University of Science and Technology
Kraków
Poland

Bogdan Rębiasz
AGH University of Science and Technology
Kraków
Poland

Janusz Opiła
AGH University of Science and Technology
Kraków
Poland

Bartłomiej Gaweł
AGH University of Science and Technology
Kraków
Poland

Tomasz Pełech-Pilichowski
AGH University of Science and Technology
Kraków
Poland

Beata Basiura
AGH University of Science and Technology
Kraków
Poland

ISSN 1434-9922 ISSN 1860-0808 (electronic)
Studies in Fuzziness and Soft Computing
ISBN 978-3-319-26492-9 ISBN 978-3-319-26494-3 (eBook)
DOI 10.1007/978-3-319-26494-3

Library of Congress Control Number: 2015954593

Springer Cham Heidelberg New York Dordrecht London
© Springer International Publishing Switzerland 2015
This work is subject to copyright. All rights are reserved by the Publisher, whether the whole or part of the material is concerned, specifically the rights of translation, reprinting, reuse of illustrations, recitation, broadcasting, reproduction on microfilms or in any other physical way, and transmission or information storage and retrieval, electronic adaptation, computer software, or by similar or dissimilar methodology now known or hereafter developed.
The use of general descriptive names, registered names, trademarks, service marks, etc. in this publication does not imply, even in the absence of a specific statement, that such names are exempt from the relevant protective laws and regulations and therefore free for general use.
The publisher, the authors and the editors are safe to assume that the advice and information in this book are believed to be true and accurate at the date of publication. Neither the publisher nor the authors or the editors give a warranty, express or implied, with respect to the material contained herein or for any errors or omissions that may have been made.

Printed on acid-free paper

Springer International Publishing AG Switzerland is part of Springer Science+Business Media
(www.springer.com)

Preface

When dealing with real-world problems, one can rarely avoid uncertainty. Very often in practice, it is not possible to obtain precise information about the values of the parameters of a modelled system. For a long time, probability approaches were dominant in the literature devoted to problems involving uncertainty. However, the most common situation in practice is when some model parameters are subject to *aleatory uncertainty*, whereas others are subject to *epistemic uncertainty*. This situation is especially common in economic calculus, which stems from the fact that economic data usually comes from various sources, but also concerns areas such as human decision making, risk analysis or engineering applications.

Aleatory uncertainty, also called *variability* or *statistical uncertainty*, is a description of naturally random behaviour in a physical process or property. Probability theory is a natural model for this type of uncertainty, and the quantification for the aleatory uncertainty is usually performed using the Monte Carlo techniques. On the other hand, epistemic uncertainty refers to limited knowledge about the system (modelled or real) or lack of information. This type of uncertainty can be reduced by, e.g. taking more measurements, conducting more tests, "buying" more information, etc. Because of that, the epistemic uncertainty is also called *reducible uncertainty*, *incertitude* or *subjective uncertainty*. Often, uncertainty quantification intends to work toward reducing epistemic uncertainties to aleatory uncertainties. However, epistemic uncertainty is not well characterised by probabilistic approaches. To evaluate epistemic uncertainties, methods such as *interval analysis*, *fuzzy logic* or *evidence theory* (Dempster–Shafer theory) are more suitable.

Measurement errors are one of the possible sources of epistemic uncertainty. It is well known that at the empirical level, uncertainty is an inseparable companion of almost any measurement. The difference between the measured and the actual values is called a *measurement error*. Since the absolute value of the measurement error can usually be bounded, it is therefore guaranteed that the actual (unknown) value of the desired quantity belongs to the *interval* (*number*) with the midpoint being the measured value, and the radius being the upper bound for the (absolute value) of the possible errors. Therefore imprecision, approximation, or uncertainty in the knowledge of the exact values of physical, technical or economic parameters can be modelled conveniently by intervals. By its nature, interval arithmetic yields rigorous enclosures for the range of operations and functions. The results are intervals in which the exact results must lie. In order that this inclusion remains true in numerical computations, the problem of round-off errors must be taken into account during the implementation of the interval computations. Proper handling of outward rounding in numerical computations forces the result to be the interval approximation of the correct real interval (that could be hypothetically obtained assuming that infinite precision is available).

A finite number of increasingly precise measurements give a finite family of usually increasingly narrower intervals (interval numbers). By assigning a respective *level of possibility* to each interval number from this family, a discrete collection of the so-called α–cuts is obtained, which can be viewed as a finite representation of a *fuzzy number*. The strict relation between interval and fuzzy numbers is often emphasised in the literature devoted to fuzzy theory. Fuzzy numbers enhance the expressive power of intervals, and therefore they are often referred to as generalised intervals.

This book presents some advances in fuzzy decision making. It is organised in eight chapters. Chapter 1 introduces some basic concepts from the fuzzy numbers theory. The main part of this chapter concerns the problem of performing arithmetic operations on fuzzy numbers. Linear operations such as addition and subtraction are rather obvious, whereas nonlinear operations, such as multiplication or division, pose a problem. Nonlinear operations usually result in fuzzy numbers of different types than operands. Therefore, various, the so-called "shape preserving", approaches to multiplication and division of fuzzy numbers are proposed in the literature. On the other hand, many researchers recommend to use the α–cuts based approach to performing operations on fuzzy numbers, because this approach allows fuzzy and interval techniques to be combine and used to effectively solve problems involving both types of uncertainty. The remaining part of this chapter is devoted to the problem of performing arithmetic operations on interactive fuzzy numbers. The reason is that all of the above-mentioned approaches implicitly assume that there is no dependency between fuzzy numbers involved in a computation. In practice this assumption is rarely satisfied, especially when dealing with economic problems. To cope with the dependency problem, stochastic simulation of fuzzy systems and nonlinear programming approaches are proposed.

The problem of comparing and ordering fuzzy numbers is described in Chap. 2. Theoretically, fuzzy numbers can only be partially ordered, and hence cannot be compared. However, in practical applications, such as decision making, scheduling, market analysis or optimisation with fuzzy uncertainties, the comparison of fuzzy numbers becomes crucial. That is why several methods for comparing and ordering fuzzy numbers have been proposed in the literature. They can be generally divided into two groups. The first group consists of methods which enable two fuzzy numbers to be compared. One can mention the probabilistic approach, centroid point approach or radius of gyration approach. To order a set of fuzzy numbers using these methods, some dedicated procedures are required. The second group consists of methods, which assign to a fuzzy number a real value. These are, for example Yager ranking index based approach, defuzzification approach or weighted average. The methods from the second group can be directly used to order a set of fuzzy numbers, by employing one of the several methods for ordering (sorting) real numbers. All the described methods are compared using a simple example.

Chapter 3 presents the concept of a fuzzy random variable and the Dempster–Shafer theory of evidence. Fuzzy random variable extends the classical definition of a random variable and is one of the possible ways to jointly consider randomness and imprecision. The simultaneous occurrence of randomness and imprecision is often the case in real-world decision problems, because data in such problems usually comes from various sources, such as historical datasets or experts opinions. The theory of evidence (also called the theory of belief functions), on the other hand, provides mathematical tools to process information, which is, at the same time, of random and imprecise nature. It allows imprecision and variability to be treated separately within a single framework. The evidence theory encompasses both possibility and probability theories.

Chapter 4 discusses selected issues of the fuzzy multi-criteria decision making (FMCDM). Generally, multi-attribute decision making (MADM) is concerned with ranking alternatives with respect to multiple criteria. Two basic techniques of multi-criteria decision making are analytical hierarchy process (AHP) and technique for order of preference by similarity to ideal solution (TOPSIS). Both AHP and TOPSIS were initially designed to deal only with crisp numbers. Later, fuzzy variants of those methods were developed, because in the real world available data are often imprecise and vague. The integrated fuzzy approach to solve multi-attribute decision problems is proposed in this chapter. Its use is illustrated using a real case from a steel industry.

Chapter 5 is devoted to a method which is able to process hybrid data, i.e. to jointly handle both randomness and imprecision. Random variables are described by probability distributions and imprecise values are modelled using possibility distributions. The main advantage of the proposed method is that it takes into account the dependencies between economic parameters. The correlation matrix is used to model dependencies between stochastic parameters, whereas interval regression is used to model both dependencies between fuzzy parameters and between fuzzy and stochastic parameters. The proposed method combines tools such as Monte Carlo simulation, interval regression and nonlinear programming. As the result of hybrid data processing, a random fuzzy set is received. Assessment of risk is obtained by computing the standard deviation and also by estimating the upper and lower cumulative distribution functions of the analysed variable. The method is verified through computing the operating profit for a metallurgical industry enterprise.

Chapter 6 describes the application of fuzzy sets to planning and scheduling of production in steel industry. Primarily the problem of steel grade assignment to customers' orders is analysed, which is the first stage of steel production planning.

Fuzzy sets are used to reduce the variety of potential steel grades and to describe the characteristic of materials by decision makers. Next, the use of fuzzy logic systems for steel production scheduling is examined. Parameters like timeliness, amount, priority and the sequence length on casters are expressed using linguistic variables. Whereas fuzzy rules are used to determine an initial schedule and to perform a quick rescheduling. More advanced systems use a multi-agent approach. Each agent may use its own fuzzy logic in order to satisfy the constraints related to the certain level of steel production. Finally, an example of a fuzzy scheduling agent is provided. It uses a genetic algorithm to generate feasible and economically efficient schedules for a continuous caster.

Chapter 7 presents a new method for forecasting the level and structure of market demand for industrial goods. The method employs two data mining methods: k-means clustering and fuzzy decision trees. The k-means method serves to separate groups with items of a similar consumption level and structure of the analysed products (consumption patterns). Whereas fuzzy decision trees are used to determine the dependencies between consumption patterns and predictors (parameters determining the level and structure of consumption). The proposed method is verified using the extensive statistical material on the level and structure of steel products consumption in selected countries during 1960–2010.

Chapter 8 discusses various techniques of visualisation of fuzzy numbers in one-, two- and more-dimensional spaces. As canonical box-and-whiskers representation is not suitable for visualisation of uncertainty in three-dimensional spaces, an approach based on ScPovPlot3D templates for POVRay, which is a powerful photorealistic renderer equipped with domain-specific programming language, dubbed scene description language (SDL), is proposed. In order to show the usefulness of the proposed technique some examples of visualisation of fuzzy objects are included, in one, two and three dimensions. For three-dimensional approach two examples of application of fuzzy visualisation are presented. The first one shows a surface defined by a function which assigns a fuzzy number to a point on the real plane. In the second example, a lesion deposited in a segment of cardiac vessel is depicted using “thick surface” approach.

Contents

Chapter 1
Fuzzy Numbers

Abstract This chapter introduces some basic concepts from the fuzzy numbers theory. The main part of the chapter concerns the problem of performing arithmetic operations on independent fuzzy numbers. The remaining part is devoted to the problem of performing arithmetic operations on interactive fuzzy numbers. To cope with the dependency problem, stochastic simulation of fuzzy systems and non-linear programming approaches are proposed.

Fuzzy set theory was introduced in the 60s by Lofti Zadeh [1]. It was initially intended to be an extension of a dual logic and/or classical set theory. However, for the last decades, it has been developed as a powerful "fuzzy" mathematics [1–4] since almost all mathematical objects can be described by sets (e.g., a function can be described by an ordered set of points). Fuzzy sets provide a mathematical framework for the precise and rigorous study of vague conceptual phenomena. Applications of fuzzy set theory can be found, for example, in computer science, artificial intelligence, control engineering, communication, medicine, decision theory, expert systems, logic, management science, pattern recognition, operations research, and robotics. The fuzzy concepts of fuzzy set theory take into account the fact that all phenomena in the physical universe have a degree of inherent uncertainty.

Real fuzzy numbers, which are of the main concern in this book, are a special case of convex, normalised fuzzy sets of the real line $\mathfrak{R}$. Fuzzy numbers play an important role in human thinking and provide a natural way of dealing with real world problems. This is because the way that people perceive the world is continually changing and cannot always be defined in true or false statements. For example, it is more natural to say "she is about eighteen" or "he is rather tall" rather than "she is eighteen year old" or "he is 185 cm tall". Fuzzy numbers are actively used in various fields, such as artificial intelligence, computer science, medicine, control engineering, decision theory, expert systems, logic, management science, operations research, pattern recognition, robotics etc. They can also be considered as a modelling language, well suited for situations involving fuzzy relations, criteria and phenomena.

© Springer International Publishing Switzerland 2015
I. Skalna et al., *Advances in Fuzzy Decision Making*,
Studies in Fuzziness and Soft Computing 333,
DOI 10.1007/978-3-319-26494-3_1

1.1 Preliminary Theory

In classical set theory each item either belongs to a set or does not belong to that set. Whereas a *fuzzy set* is as a class of objects with continuum of grades of membership (characteristic) function (see, e.g., [1, 3, 5]). The following example pictures the difference between a classical and a fuzzy set.

Example 1.1 Consider the set of "tall men". In classical set theory the statement "people taller than or equal to 1.82 m are tall" can be represented graphically by a step function as depicted in Fig. 1.1. This crisp membership function does not seem to work well, since it makes no distinction between somebody who is 1.86 m and someone who is 2.16 m, they are both simply tall. Moreover, the difference between a 1.80 and 1.83 man is only 3 cm, however the membership function just says one is tall and the other is not tall. The fuzzy set approach to the set of tall men provides a much better representation of the tallness of a person. The set, shown in Fig. 1.2, is defined by a continuously increasing function. So, a person with the grade of membership equal to 0.95 is a really tall person, whereas a person with the grade of membership equal to 0.3 is really not very tall at all.

Definition 1.1 A fuzzy set $\tilde{A}$ on the domain X is a pair $(X, \mu_{\tilde{A}})$, where $\mu_{\tilde{A}} : X \to [0, 1]$ is called the grade of membership of x in $(X, \mu_{\tilde{A}})$.

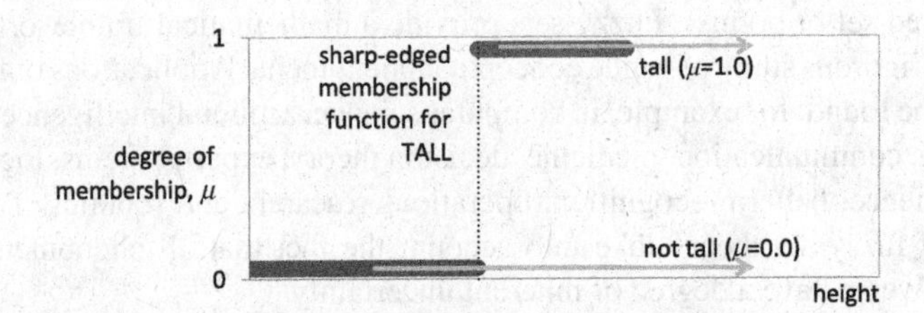

Fig. 1.1 Crisp membership function for the set of tall men

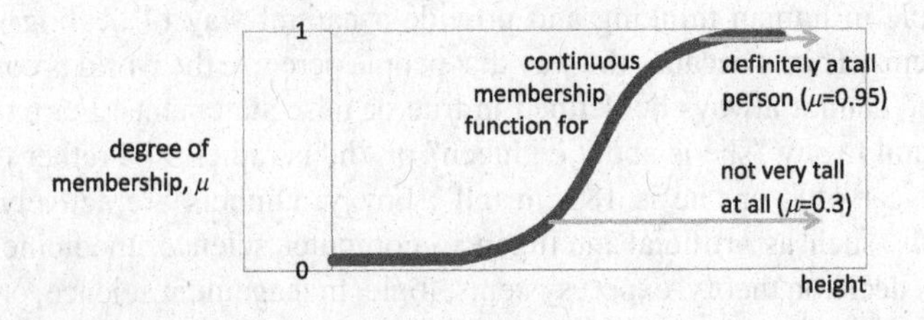

Fig. 1.2 Fuzzy membership function for the set of tall men

A fuzzy set $\tilde{A}$ can also be expressed as a set of ordered pairs. The following two notations are used the most often:

$$\tilde{A} = \left\{(x, \mu_{\tilde{A}}(x)) \mid x \in X, \mu_{\tilde{A}} : X \to [0, 1]\right\}, \tag{1.1}$$

$$\tilde{A} = \left\{\langle(x, \mu_{\tilde{A}}(x))\rangle \mid x \in X, \mu_{\tilde{A}} : X \to [0, 1]\right\}. \tag{1.2}$$

A fuzzy set $(X, \mu_{\tilde{A}})$ with a finite domain is usually denoted by $\{\mu_{\tilde{A}}(x_1)/x_1, \ldots, \mu_{\tilde{A}}(x_n)/x_n\}$ or $\left\{\frac{\mu_{\tilde{A}}(x_1)}{x_1}, \ldots, \frac{\mu_{\tilde{A}}(x_n)}{x_n}\right\}$.

A *fuzzy number* is a special case of a convex, normalised fuzzy set of the real line. Complex fuzzy numbers can also be encountered in the literature, however, they are out of the scope of this book.

Definition 1.2 A real fuzzy number $\tilde{A}$ is a fuzzy subset of the real numbers $\Re$ characterised by a membership function $\mu_{\tilde{A}} : \Re \to [0, 1]$ that assigns to each $x \in \Re$ a grade of membership $\mu_{\tilde{A}}(x)$. It is required that the membership function fulfils the following four conditions:

1. *Normality*: there exists $x_0 \in \Re$, such that $\mu_{\tilde{A}}(x_0) = 1$.
2. *Convexity*: $\forall x, y \in \Re, \forall \lambda \in \Re$

$$\mu_{\tilde{A}}(\lambda x + (1 - \lambda)y) \geqslant \min\{\mu_{\tilde{A}}(x), \mu_{\tilde{A}}(y)\}. \tag{1.3}$$

3. *Upper semi-continuity* (u.s.c., see Fig. 1.3a): for all $x_0 \in \Re$ and all $\varepsilon > 0$ there exist a neighbourhood $V(x_0)$, such that for all $x \in V(x_0)$

$$\mu_{\tilde{A}}(x_0) \leqslant \mu_{\tilde{A}}(x) + \varepsilon. \tag{1.4}$$

4. *Compactness*: the support of a fuzzy number

$$supp(\tilde{A}) = cl\{x \in \Re \mid \mu_{\tilde{A}}(x) > 0\} \tag{1.5}$$

is bounded; $cl(S)$ denotes the closure of a set S, i.e., the set S together with all of its limit points.

The set of all real fuzzy numbers will be denoted throughout by $\mathcal{F}(\Re)$.

Definition 1.3 The core of a fuzzy number $\tilde{A}$ (see Fig. 1.3a) is defined as a set of all points for which the value of the membership function equals 1, i.e.,

$$core(\tilde{A}) = \{x \in \Re \mid \mu_{\tilde{A}}(x) = 1\}. \tag{1.6}$$

Definition 1.4 Let $\tilde{A} \in \mathcal{F}(\Re)$ and let $\alpha \in (0, 1]$. Then, an α-cut of $\tilde{A}$ is defined as the following closed interval:

$$\tilde{A}^{\alpha} = \{x \in \Re \mid \mu_{\tilde{A}}(x) \geqslant \alpha\}. \tag{1.7}$$

The 0-cut is defined separately (otherwise it would give the entire real line) and equals to the support of a fuzzy number, $\tilde{A}^0 = \text{supp}(\tilde{A})$. The A^1–cut is simply the core of a fuzzy number $\tilde{A}$. A strong (or strict) α-cut is an open interval defined by $\tilde{A}^\alpha = \{x \in \Re \mid \mu_{\tilde{A}}(x) > \alpha\}$.

The well-known properties (consistency conditions) of α–cuts are the following: for $\alpha, \beta \in [0, 1]$

- $\tilde{A}^\alpha \subseteq \tilde{A}^\beta$ for $\alpha \geqslant \beta$ (monotonicity),
- $\tilde{A}^\alpha = \bigcap_{\beta<\alpha} \tilde{A}^\beta$ (continuity).

A fuzzy number $\tilde{A}$ can be represented by an infinite family of nested α–cuts (see Fig. 1.3b)

$$\tilde{A} = \bigcup_{\alpha\in[0,1]} (\alpha, \tilde{A}^\alpha). \tag{1.8}$$

This representation is referred to as *parametric representation* of a fuzzy number and is extensively used in various methods for solving problems involving fuzzy numbers. In practical computations it is recommended to select a finite subset of α–cuts with relevant degrees of membership; they must also be semantically distinguishable. The problem of selecting the best approximation of a fuzzy number by a finite family of its α–cuts was considered, e.g., by Pedrycz [6].

Theorem 1.1 (Representation Theorem) *Let* $\tilde{A} \in \mathcal{F}(\Re)$. *Then,*

$$\mu_{\tilde{A}}(x) = \sup_{\alpha\in[0,1]} \left\{\min(\alpha, \mu_{\tilde{A}^\alpha}(x))\right\}, \tag{1.9}$$

where

$$\mu_{\tilde{A}^\alpha}(x) = \begin{cases} 1, \; x \in \tilde{A}^\alpha, \\ 0, \text{ otherwise.} \end{cases} \tag{1.10}$$

The concept of α–cuts plays an essential role in the fuzzy numbers theory. The α–cuts representation is often used to process fuzzy data by means of interval (*level-by-level*) techniques. This allows to combine fuzzy and interval techniques and to effectively solve problems involving both types of uncertainty.

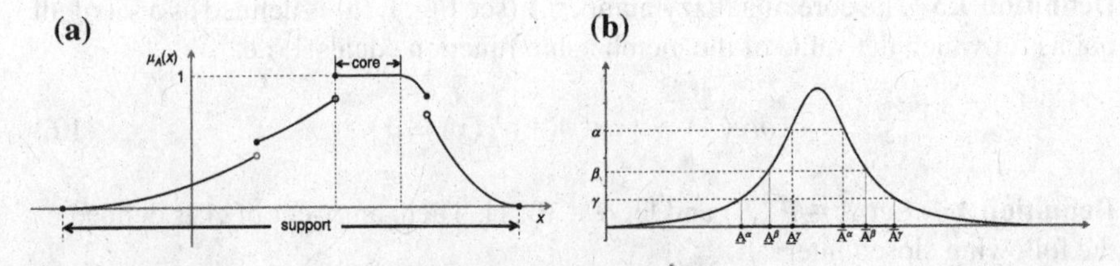

Fig. 1.3 General form of a fuzzy quantity with upper semi-continuous membership function (**a**); nested α–cuts corresponding to the selected levels: α–level, β–level and γ–level (**b**)

1.1.1 LR–type Fuzzy Numbers

LR-type fuzzy numbers are an important subclass of fuzzy numbers [7]. The solution of fuzzy problems is somewhat easier if only LR–type fuzzy numbers are involved.

Definition 1.5 An LR-type fuzzy number $\tilde{A}$ is described by the following membership function [7]:

$$\mu_{\tilde{A}}(x) = \begin{cases} L_{\tilde{A}}\left(\frac{a-x}{\alpha}\right), & x \in [a-\alpha, a], \\ 1, & x \in [a, b], a \leqslant b, \\ R_{\tilde{A}}\left(\frac{x-b}{\beta}\right), & x \in [b, b+\beta], \\ 0, & \text{otherwise.} \end{cases} \tag{1.11}$$

where a and b are, respectively, the lower and upper *modal values*, $[a, b]$ is the core of $\tilde{A}$, $\alpha > 0$ is the left spread, $\beta > 0$ is the right spread, and $L_{\tilde{A}}$, $R_{\tilde{A}}$ are the so-called shape functions, $L_{\tilde{A}}, R_{\tilde{A}} : [0, 1] \to [0, 1]$, with $L_{\tilde{A}}(0) = R_{\tilde{A}}(0) = 1$ and $L_{\tilde{A}}(1) = R_{\tilde{A}}(1) = 0$.

The $L_{\tilde{A}}$ and $R_{\tilde{A}}$ functions are non-increasing, continuous mappings [7]. The support of an *LR*-type fuzzy number $\text{supp}(\tilde{A}) = [a-\alpha, b+\beta]$. An *LR*-type fuzzy number is usually denoted by $\tilde{A} = (a, b, \alpha, \beta)_{LR}$. However, if $L_{\tilde{A}}(x) = R_{\tilde{A}}(x) = 1 - x$, then a simpler notation, $\tilde{A} = (a, b, \alpha, \beta)$, is usually used. The set of all LR-type fuzzy numbers will be denoted throughout by $\mathcal{F}(\Re)_{LR}$.

Two types of *LR*-type fuzzy numbers are the most commonly used in practice. These are *triangular* (TFN) (Fig. 1.4) and *trapezoidal* (TRFN) (Fig. 1.5) fuzzy numbers with both shape functions linear.

Definition 1.6 A triangular fuzzy number is defined by the following triangle-shaped membership function:

$$\mu_{\tilde{A}}(x) = \begin{cases} 0, & x \leqslant a, \\ \frac{x-a}{b-a}, & a < x \leqslant b, \\ \frac{c-x}{c-b}, & b \leqslant x < c, \\ 0, & x \geqslant c. \end{cases} \tag{1.12}$$

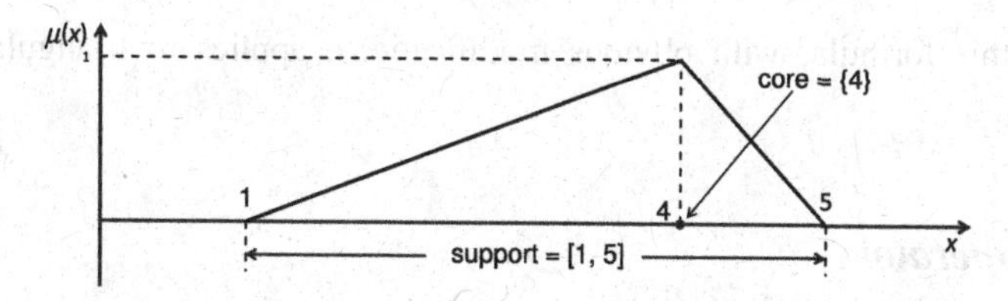

Fig. 1.4 A triangular fuzzy number $\tilde{A} = (1, 4, 5)$ with $\ker(\tilde{A}) = \{4\}$ and $\text{supp}(\tilde{A}) = [1, 5]$

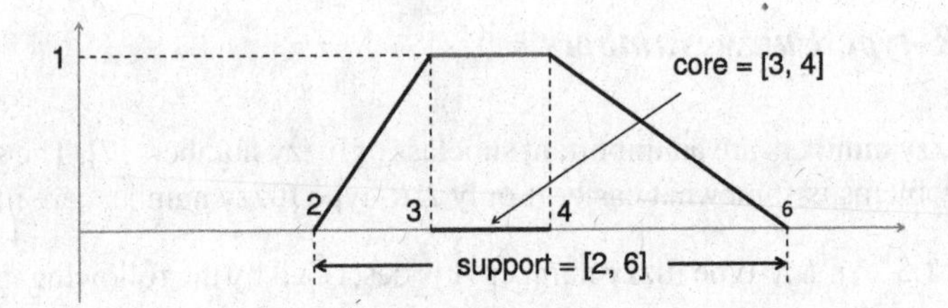

Fig. 1.5 A trapezoidal fuzzy number $\tilde{A} = (2, 3, 4, 6)$ with $\ker(\tilde{A}) = [3, 4]$ and $\text{supp}(\tilde{A}) = [2, 6]$

The above definition differs from the general definition of *LR*-type fuzzy numbers. This stems from that it is adopted to the notation of triangular fuzzy number used in this book, which is the following: $\tilde{A} = (a, b, c)$, where $a = \inf(\text{supp}(\tilde{A}))$, $c = \sup(\text{supp}(\tilde{A}))$ and b is a modal (central) value. A similar notation will be used for trapezoidal fuzzy numbers.

Trapezoidal fuzzy numbers are often called fuzzy intervals. They form the most general class of fuzzy numbers with linear shape functions.

Definition 1.7 A trapezoidal fuzzy numbers is defined by the following trapeze-shaped membership function:

$$\mu_{\tilde{A}}(x) = \begin{cases} 0, & x \leqslant a, \\ \frac{x-a}{b-a}, & a < x \leqslant b, \\ 1, & b < x < c, \\ \frac{d-x}{d-c}, & c \leqslant x < d, \\ 0, & x \geq d. \end{cases} \tag{1.13}$$

A trapezoidal fuzzy number will be denoted as $\tilde{A} = (a, b, c, d)$, where $a = \inf(\text{supp}(\tilde{A}))$, $d = \sup(\text{supp}(\tilde{A}))$, and b and c are, respectively, lower and upper modal values. Note that trapezoidal fuzzy numbers are a generalisation of triangular fuzzy numbers.

The α–cuts of a trapezoidal fuzzy number can be presented in the form of intervals with endpoints being linear functions of α. Given a trapezoidal fuzzy number $\tilde{A} = (a, b, c, d)$, its α–cuts are given by:

$$\tilde{A}^{\alpha} = [(b - a)\alpha + a, d - (d - c)\alpha]. \tag{1.14}$$

Note that this formula, with obvious modifications, applies to triangular fuzzy numbers.

1.1.2 T-Operators

The triangular norms (t-norms), triangular conorms (t-conorms) an important role in fuzzy set theory. There is a large amount of literature devoted to t-norm based operations on fuzzy numbers. The t-norms and t-conorms generalise, respectively, the conjunctive ("AND") and disjunctive ("OR") operators, and hence can be used to define the intersection and union operations in fuzzy logic and fuzzy inference systems. This possibility was first noted by Höhle [8] and was later exploited by Alsina et al. [9], Klement [10] and Dubois and Prade [11]. The properties of t-norms for possible use in the development of intelligent systems were investigated, e.g., by Bonissone [12]. T-operators has also been widely used in the design of fuzzy logic controllers, in the modelling of other decision-making processes [13, 14] and in the scheduling of production. The current notion of t-norms is due to Schweizer and Sklar [15].

Definition 1.8 A function $T : [0, 1] \times [0, 1] \to [0, 1]$ is a t-norm if and only if for any $x, y, z \in [0, 1]$

(i) $T(x, 1) = x$ (existence of a unit 1),
(ii) $x \leqslant y \Rightarrow T(x, z) \leqslant T(y, z)$ (monotonicity),
(iii) $T(x, y) = T(y, x)$ (commutativity),
(iv) $T(x, T(y, z)) = T(T(x, y), z)$ (associativity).

A t-norm is Archimedean, iff:

(i) $T(x, y)$ is continuous,
(ii) $T(x, x) < x \; \forall x \in (0, 1)$.

An Archimedean t-norm is strict, iff

$$T(x', y') < T(x, y), \text{ if } x' < x, y' < y, \forall x', y', x, y \in (0, 1).$$

The most prominent examples of t-norms are the following:

- *Minimum t-norm* (also called min-norm, Gödel t-norm or standard t-norm):

$$T_{\min}(x, y) = \min\{x, y\}.$$

- *Standard product or probabilistic t-norm* (the ordinary product of real numbers)

$$T_{prod}(x, y) = x * y,$$

- *Lukasiewicz t-norm (also called bounded difference)*

$$T_{Luk}(x, y) = \max\{0, x + y - 1\}.$$

- *Drastic t-norm (also called weak t-norm)*

$$T_d(x, y) = \begin{cases} y, \text{ if } x = 1, \\ x, \text{ if } y = 1, \\ 0, \text{ otherwise.} \end{cases}$$

This is the only one t-norm which is not continuous.

- Hamacher t-norm

$$T_H(x, y) = \frac{\lambda xy}{1 - (1 - \lambda)(x + y - xy)}.$$

The drastic t-norm is the weakest t-norm, and the minimum t-norm is the strongest one, i.e., $T_d(x, y) \leqslant T(x, y) \leqslant T_{\min}(x, y)$, for any t-norm T and all $x, y \in [0, 1]$.

Definition 1.9 A function $\perp : [0, 1] \times [0, 1] \to [0, 1]$ is a t-conorm if and only if for any $x, y, z \in [0, 1]$

(i) $\perp(x, 0) = x$ (existence of a zero 0),
(ii) $x \leqslant y \Rightarrow \perp(x, z) \leqslant \perp(y, z)$ (monotonicity),
(iii) $\perp(x, y) = \perp(y, x)$ (commutativity),
(iv) $\perp(x, \perp(y, z)) = \perp(\perp(x, y), z)$ (associativity).

A t-conorm is Archimedean, iff:

(i) $\perp(x, y)$ is continuous,
(ii) $\perp(x, x) > x \; \forall x \in (0, 1)$.

An Archimedean t-conorm is strict, iff

$$\perp(x', y') < \perp(x, y), \text{ if } x' < x, y' < y, \forall x', y', x, y \in (0, 1).$$

The most popular t-conorms are listed below.

- *Maximum t-conorm* (also called min-norm, Gödel t-norm or standard t-norm):

$$\perp_{\max}(x, y) = \max\{x, y\},$$

- *Probabilistic t-conorm*

$$\perp_{prod}(x, y) = x + y - x * y,$$

- *Lukasiewicz t-conorm*

$$\perp_{Luk}(x, y) = \min\{1, x + y\},$$

- *Drastic t-conorm (also called a weak t-conorm)*

$$\perp_d(x, y) = \begin{cases} y, & \text{if } x = 0, \\ x, & \text{if } y = 0, \\ 0, & \text{otherwise.} \end{cases}$$

This is the only one t-conorm which is not continuous.

- Hamacher t-norm

$$\perp_H(x, y) = \frac{\lambda(x + y) + xy(1 - 2\lambda)}{\lambda + xy(1 - \lambda)}.$$

Dually to t-norms, all t-conorms are bounded by the maximum and the drastic t-conorm, i.e., $\perp_{\max}(x, y) \leq \perp(x, y) \leq \perp_{\mathrm{d}}(x, y)$, for any t-conorm $\perp$ and all $x, y \in [0, 1]$.

1.2 Arithmetic of Fuzzy Numbers

The fuzzy arithmetic allows mathematical operators such as addition, subtraction, multiplication, and division, to be applied to the fuzzy domain, i.e., it is able to propagate the fuzzy uncertainty through computations. Different approaches to performing these operations, yielding sometimes different results, can be found in, e.g., Dubois and Prade [16], Filev and Yager [17], Kreinovich and Pedrycz [18], Fodor and Bede [19] and Tsao [20]. The comparison of four methods for multiplication of fuzzy numbers was presented, e.g., by Zhang et al. [21] and Dutta et al. [22]. They recommended the α–cut based approach for executing fuzzy arithmetic. The α–cut based approach employs interval arithmetic which additionally emphasizes the relation between fuzzy and interval numbers. Bansal [23] presented the basic mathematical operations formulated on trapezoidal fuzzy numbers. Roy [24] described arithmetic operations for general trapezoidal fuzzy numbers. Extension of algebraic mathematics for positive trapezoidal fuzzy numbers was described, e.g., by Vahidi and Rezvani [25]. Kolesárová [26] defined the t-norm based multiplication of *LR*-type fuzzy numbers with positive supports. Dubois and Prade [27] proposed an approximate formula for multiplications of fuzzy number with positive supports. Nevertheless, regardless which approach is used, the algebraic system of fuzzy arithmetic is only an *abelian monoid* under both addition and multiplication, additive and multiplicative inverses exist only for fuzzy numbers with support of zero width. In general case, the difference between two same fuzzy numbers results in a fuzzy number with support symmetrical around the modal value which is equal to zero (such fuzzy number is sometimes considered as fuzzy zero). Several other algebraic laws, valid for real numbers, are not preserved in fuzzy arithmetic.

1.2.1 Zadeh's Extension Principle

The most general definition of operations on fuzzy numbers is based on the so-called *extension principle*. It was first introduced by Zadeh [1], that is why it is usually called *Zadeh's extension principle* [1]. Later on, several modifications were suggested (see, e.g., [28]).

The most general form of extension principle is called the t-norm (triangular norm) based extension principle [7].

Definition 1.10 (*Extension principle*) Let the mapping $f : X \times Y \to Z$, where $X, Y, Z \subseteq \Re$, be given $\Re$. It can be extended to fuzzy numbers in the following way:

$$\mu_{f(\tilde{A},\tilde{B})}(z) = \sup_{z=f(x,y)} T(\mu_{\tilde{A}}(x), \mu_{\tilde{B}}(y)), \tag{1.15}$$

where $\tilde{A} \in \mathcal{F}(X)$, $\tilde{B} \in \mathcal{F}(Y)$ and T is a t-norm.

For a binary arithmetic operation, Zadeh's extension principle can be defined as follows.

Definition 1.11 Given two fuzzy numbers $\tilde{A}$, $\tilde{B}$ and an arithmetic operation $\circ \in \{+, -, *, /\}$ (in the case of division, it is assumed that $0 \notin \text{supp}(\tilde{B})$), the result of an operation $\tilde{C} = \tilde{A} \circ \tilde{B}$ is defined by the following membership function:

$$\mu_{\tilde{C}}(z) = \sup_{z=x\circ y} T(\mu_{\tilde{A}}(x), \mu_{\tilde{B}}(y)). \tag{1.16}$$

The most frequently used t-norm is the strongest min-norm, however, extension principle with other norms can also be encountered in the literature. The min-norm based Zadeh's extension principle takes the form:

$$\mu_{\tilde{A}\circ\tilde{B}}(z) = \sup_{z=x\circ y} \min(\mu_{\tilde{A}}(x), \mu_{\tilde{B}}(y)). \tag{1.17}$$

Example 1.2 Consider two trapezoidal fuzzy numbers $\tilde{A} = (-2, -1, 3, 8)$ and $\tilde{B} = (-4, 1, 5, 6)$, and take the corresponding discrete fuzzy sets

$$\tilde{A}' = \{0/-2, 1/-1, 1/0, 1/1, 1/2, 1/3, 0.8/4, 0.6/5, 0.4/6, 0.2/7, 0/8\},$$
$$\tilde{B}' = \{0/-4, 0.2/-3, 0.4/-2, 0.6/-1, 0.8/0, 1/1, 1/2, 1/3, 1/4, 1/5, 0/6\}.$$

The goal is to find the membership $\mu_{\tilde{A}'+\tilde{B}'}$. Table 1.1 shows all possible results of addition of the elements from the domains of $\tilde{A}'$ and $\tilde{B}'$. The corresponding operands and their membership grades are given on the top and on the left side of the table. The final result obtained using Zadeh's extension principle (1.17) is depicted in Fig. 1.6.

1.2.2 α–cuts Based Operations

Zadeh's extension principle, in spite of all its usefulness, has been regarded as a time-consuming and expensive computational tool since its usage requires at each sample the inverse function in addition to the effect induced by the discretization process. That is why, the α–cuts based approach is often used instead.

Table 1.1 The sums of the elements from the domains of $\tilde{A}'$ and $\tilde{B}'$

		0	0.2	0.4	0.6	0.8	1	1	1	1	1	0
		−4	−3	−2	−1	0	1	2	3	4	5	6
0	−2	−6	−5	−4	−3	−2	−1	0	1	2	3	4
1	−1	−5	−4	−3	−2	−1	0	1	2	3	4	5
1	0	−4	−3	−2	−1	0	1	2	3	4	5	6
1	1	−3	−2	−1	0	1	2	3	4	5	6	7
1	2	−2	−1	0	1	2	3	4	5	6	7	8
1	3	−1	0	1	2	3	4	5	6	7	8	9
0.8	4	0	1	2	3	4	5	6	7	8	9	10
0.6	5	1	2	3	4	5	6	7	8	9	10	11
0.4	6	2	3	4	5	6	7	8	9	10	11	12
0.2	7	3	4	5	6	7	8	9	10	11	12	13
0	8	4	5	6	7	8	9	10	11	12	13	14

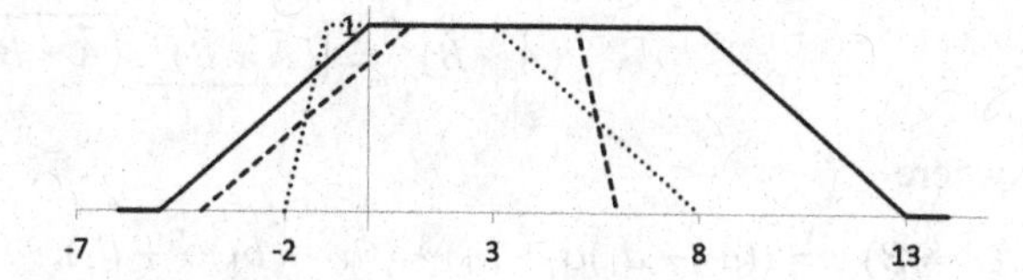

Fig. 1.6 Discrete fuzzy numbers $\tilde{A}'$, $\tilde{B}'$ and their sum obtained using Zadeh' extension principle

Definition 1.12 Given two fuzzy numbers $\tilde{A}$, $\tilde{B}$ and an arithmetic operation $\circ \in \{+, -, *, /\}$, the result is defined by the following α–cuts:

$$(\tilde{A} \circ \tilde{B})^{\alpha} = \tilde{A}^{\alpha} \circ \tilde{B}^{\alpha} = \{x \circ y | x \in \tilde{A}^{\alpha}, y \in \tilde{B}^{\alpha}\}, \ \alpha \in [0, 1]. \tag{1.18}$$

The result of an arithmetic operation $(\tilde{A} \circ \tilde{B}) = \bigcup_{\alpha \in [0,1]} \alpha (\tilde{A} \circ \tilde{B})^{\alpha}$ is always a fuzzy number.

Using the α–cuts representation (1.14), an arithmetic operation on two trapezoidal fuzzy numbers $\tilde{A} = (a_1, a_2, a_3, a_4)$ and $\tilde{B} = (b_1, b_2, b_3, b_4)$ can be performed for each $\alpha \in [0, \ 1]$ using the following formula:

$$(\tilde{A} \circ \tilde{B})^{\alpha} = [(a_2 - a_1)\alpha + a_1, a_4 - (a_4 - a_3)\alpha] \circ [(b_2 - b_1)\alpha + b_1, b_4 - (b_4 - b_3)\alpha]. \tag{1.19}$$

In order to calculate the resulting α-cut, $(\tilde{A} \circ \tilde{B})^{\alpha}$, interval arithmetic [29] must be employed.

Linear α–cuts based operations on fuzzy numbers, such as addition and subtraction, are straightforward. For two trapezoidal fuzzy numbers $\tilde{A} = (a_1, a_2, a_3, a_4)$ and $\tilde{B} = (b_1, b_2, b_3, b_4)$, they are defined as follows:

$$\tilde{A} + \tilde{B} = (a_1 + b_1, a_2 + b_2, a_3 + b_3, a_4 + b_4), \tag{1.20}$$

$$\tilde{A} - \tilde{B} = (a_1 - b_4, a_2 - b_3, a_3 - b_2, a_4 - b_1). \tag{1.21}$$

However, in the case of nonlinear operations, such as multiplication and division, the resulting fuzzy number in general case will be of a different type than operands. For convenience and also to simplify the computation, the resulting fuzzy number can be approximated by a fuzzy number of the respective type (see Example 1.3). For triangular and trapezoidal fuzzy numbers the choice of the most appropriate approximation is obvious. Otherwise, this choice could be non-trivial.

In the case when both operands are non-negative (i.e., have non-negative supports), the α–cuts based multiplication takes the form:

$$(\tilde{A} * \tilde{B})^{\alpha} = [\underline{(\tilde{A} * \tilde{B})^{\alpha}}, \overline{(\tilde{A} * \tilde{B})^{\alpha}}], \tag{1.22}$$

where

$$\underline{(\tilde{A} * \tilde{B})^{\alpha}} = ((a_2 - a_1)\alpha + a_1) * ((b_2 - b_1)\alpha + b_1),$$
$$\overline{(\tilde{A} * \tilde{B})^{\alpha}} = (a_4 - (a_4 - a_3)\alpha) * (b_4 - (b_4 - b_3)\alpha).$$

Then, the membership function $\mu_{\tilde{A}*\tilde{B}}(z)$ can be easily calculated form (1.22), by equating both endpoints of the above interval to x.

The α–cuts based division of fuzzy numbers, providing that $0 \notin \text{supp}(\tilde{B})$, is defined as follows:

$$(\tilde{A}/\tilde{B})^{\alpha} = \tilde{A}^{\alpha} * (1/\tilde{B}^{\alpha}), \tag{1.23}$$

where $1/\tilde{B}$ is the reciprocal of a fuzzy number. In the case when $\tilde{B}$ non-negative, the reciprocal takes the form:

$$(1/\tilde{B}^{\alpha}) = [1/(b_4 - (b_4 - b_3)\alpha), 1/((b_2 - b_1)\alpha + b_1)]. \tag{1.24}$$

1.2.3 Shape-Preserving Operations

In the case of trapezoidal fuzzy numbers with positive supports, the trapezoidal result of multiplication and division can be obtained using the following formulae:

$$(\tilde{A} * \tilde{B}) = (a_1 * b_1, a_2 * b_2, a_3 * b_3, a_4 * b_4), \tag{1.25}$$

$$(\tilde{A}/\tilde{B}) = (a_1/b_4, a_2/b_3, a_3/b_2, a_4/b_1). \tag{1.26}$$

It can be shown that the support and the core of the resulting fuzzy number coincide with the support and the core of the fuzzy number obtained using the α–cuts based approach. The operations (1.25) and (1.26) are called the shape preserving operations. Another approaches to shape-preserving operations on fuzzy numbers can be found, e.g., in [30]. Given two triangular fuzzy numbers $\tilde{A} = (a_1, a_2, a_3)$ and $\tilde{B} = (b_1, b_2, b_3)$ with positive central values ($a_2 > 0, b_2 > 0$), the shape-preserving multiplication of triangular fuzzy numbers has the following form [30]:

$$\tilde{A} *_{T_d} \tilde{B} = (\min(a_1 b_2, a_2 b_1), a_2 b_2, \max(a_3 b_2, a_2 b_3)). \tag{1.27}$$

The drastic t-norm is proved to be the norm which preserves the shape of fuzzy numbers.

For triangular fuzzy numbers $\tilde{A} = (a_1, a_2, a_3)$ and $\tilde{B} = (b_1, b_2, b_3)$ with positive supports ($a_1 > 0, b_1 > 0$), the same result can be obtained using the formula proposed by Kolesárová [26]:

$$\tilde{A} *_{T_d} \tilde{B} = (a_2 b_2(1 - d_1), a_2 b_2, a_2 b_2(1 + d_2)), \tag{1.28}$$

where

$d_1 = \max((a_2 - a_1)/a_2, (b_2 - b_1)/b_2)$,
$d_2 = \max((a_3 - a_2)/a_2, (b_3 - b_2)/b_2)$.

Bansal [23] presented formulae for selected arithmetic operations as well as basic functions. For two arbitrary trapezoidal fuzzy numbers $\tilde{A} = (a_1, a_2, a_3, a_4)$ and $\tilde{B} = (b_1, b_2, b_3, b_4)$, the so-called *extended multiplication* rule has the form:

$$\tilde{A} * \tilde{B} = (d_1, d_2, d_3, d_4), \tag{1.29}$$

where

$d_1 = \min\{a_1 b_1, a_1 b_4, a_4 b_1, a_4 b_4\}$,
$d_2 = \min\{a_2 b_2, a_3 b_2, a_3 b_3, a_2 b_3\}$,
$d_3 = \max\{a_2 b_2, a_3 b_2, a_3 b_3, a_2 b_3\}$,
$d_4 = \max\{a_1 b_1, a_1 b_4, a_4 b_1, a_4 b_4\}$.

Assuming that $0 \notin [b1, b4]$, the division of two trapezoidal fuzzy numbers takes the form [23, 31]:

$$\tilde{A}/\tilde{B} = (d_1, d_2, d_3, d_4), \tag{1.30}$$

where

$d_1 = \min\{a_1/b_1, a_1/b_4, a_4/b_1, a_4/b_4\}$,
$d_2 = \min\{a_2/b_2, a_2/b_3, a_3/b_2, a_3/b_3\}$,
$d_3 = \max\{a_2/b_2, a_2/b_3, a_3/b_2, a_3/b_3\}$,
$d_4 = \max\{a_1/b_1, a_1/b_4, a_4/b_1, a_4/b_4\}$.

Dubois and Prade [27] derived the following approximate formula for the $T_{\min}$-multiplication of *LR*-type fuzzy numbers $\tilde{A} = (a_1, a_2, a_3)$ and $\tilde{B} = (b_1, b_2, b_3)$ with positive supports:

$$\tilde{A} *_{T_M} \tilde{B} \approx (a_2 b_1 + a_1 b_2 - a_2 b_2, a_2 b_2, a_2 b_3 + a_3 b_2 - a_2 b_2). \tag{1.31}$$

The above formula holds providing that $\frac{a_2}{a_2 - a_1} + \frac{b_2}{b_2 - b_1} \gg 1$. The approximation is necessary to preserve the shape of the L, R functions, since unlike addition and subtraction multiplication based on the strongest t-norm does is not preserving operation.

1.2.4 Examples

Example 1.3 Let the following two triangular fuzzy numbers be given $\tilde{A} = (1, 2, 4)$, $\tilde{B} = (1, 4, 6)$ (see Fig. 1.7a) and for clarity of presentation put $\tilde{C} = \tilde{A} * \tilde{B}$.

The results obtained using formulae (1.22), (1.25), (1.27)–(1.29) and (1.31) are presented in Fig. 1.7b. The solid black line represents the product obtained using the α–cuts based approach (1.22). The dashed grey line represents the product obtained using the formula (1.31), the dashed black line denotes the product obtained using the formulae (1.25) and (1.29), whereas the grey solid line denotes the product obtained using formulae (1.27) and (1.28). It can be seen that the obtained results have the same modal value, but they differ in the width of the supports.

The results obtained by using the formulae (1.22), (1.25) and (1.29) have the same support $\text{supp}(\tilde{C}) = [1, 24]$, which is the widest among all the supports. However, the shape functions of the fuzzy number produced by the formula (1.22) are no longer linear, they are nonlinear functions of x:

$$\mu_{\tilde{C}}(x) = \begin{cases} \frac{-2+\sqrt{3x+1}}{3}, & 1 \leqslant x < 8, \\ \frac{5-\sqrt{x+1}}{2}, & 8 \leqslant x \leqslant 24. \end{cases} \tag{1.32}$$

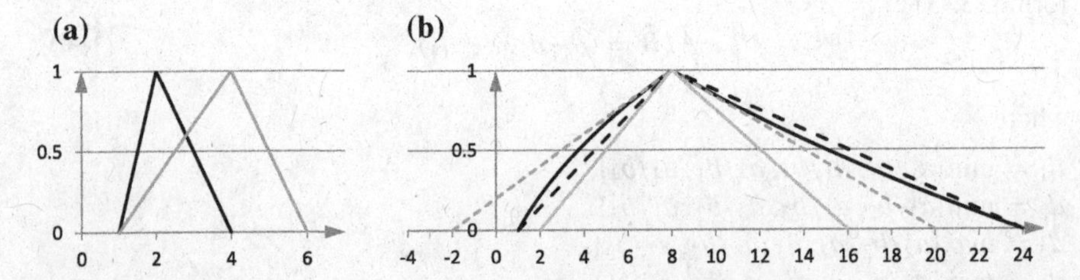

Fig. 1.7 Two exemplary triangular fuzzy numbers $\tilde{A} = (1, 2, 4)$ and $\tilde{B} = (1, 4, 6)$ (**a**); the results of multiplication of $\tilde{A}$ and $\tilde{B}$ using different multiplication formulae (**b**)

In order to remain within the class of triangular fuzzy numbers, the result of the formula (1.22) can be approximated by the triangular fuzzy number (1, 8, 24) (obtained by taking the three characteristic points: lower bound of the support, modal value and upper bound of the support), which coincides with the result of the formulae (1.25) and (1.29).

The formulae (1.27) and (1.28) give the triangular fuzzy number $\tilde{C} = (2, 8, 16)$ with the narrower support $\text{supp}(\tilde{C}) = [2, 18]$ being about 30 % narrower than the result of formulae (1.22), (1.25) and (1.29).

Finally, the approximate formula (1.31) results in the fuzzy number $\tilde{C} = (-2, 8, 20)$ with partially negative support $\text{supp}(\tilde{C}) = [-2, 20]$, which contradicts with the rather obvious expectation that the multiplication of two positive quantities should yield a positive quantity. This support is about 27 % wider than the result of the formula (1.27) and about 4 % narrower than the results of the formulae (1.22), (1.25) and (1.29). Crisp values obtained using the centroid defuzzification approach (Centre of Area, COA) approach, which is described in Sect. 1.2, are presented in Table 1.2.

Summarising, the obtained results differ significantly. This concerns both fuzzy and crisp results. Which formula to use in a specific problem remains an open question. However, many authors are inclined to use the α-cut based operations on fuzzy numbers, for the reasons described at the beginning of this chapter.

As already mentioned, the reciprocal of an arbitrary fuzzy number can be obtained using the α–cut based approach, providing the support of this number does not contain zero. When dealing with trapezoidal fuzzy number $\tilde{A} = (a_1, a_2, a_3, a_4)$ with positive (or negative) support, the following formula can be used for computing the reciprocal:

$$1/\tilde{A} = (1/a_4, 1/a_3, 1/a_2, 1/a_1). \tag{1.33}$$

It can be shown that the support and the core of the resulting trapezoidal fuzzy number coincide with the support and the core of the fuzzy number obtained using the α–cuts based approach.

In the case of triangular fuzzy numbers, the same result can be obtained using the following formula for division of triangular fuzzy numbers:

$$(1, 1, 1)/\tilde{A} = (\alpha, \beta, \gamma), \tag{1.34}$$

where
$\alpha = \min(a_1/b_1, a_1/b_3, a_3/b_1, a_3/b_3)$,
$\beta = a_2/b_2$,
$\gamma = \max(a_1/b_1, a_1/b_3, a_3/b_1, a_3/b_3)$.

Table 1.2 Crisp values obtained using the Centre of Area (COA) defuzzification method

Formula	Value
(1.22)	18.61
(1.25), (1.29)	19.00
(1.27), (1.28)	16.67
(1.31)	16.67

Example 1.4 Let the fuzzy number $\tilde{A} = (3, 5, 10)$ be given (black solid line in Fig. 1.8a). The reciprocal of $\tilde{A}$, obtained from the formula (1.34), is the fuzzy number $1/\tilde{A} = (0.1, 0.2, 1/3)$, whereas the α–cuts based approach gives the fuzzy number with the following membership function:

$$\mu_{1/\tilde{A}}(x) = \begin{cases} 2 - 0.2/x, & 0.1 \leqslant x \leqslant 0.2, \\ 0.5/x - 1.5, & 0.2 \leqslant x \leqslant 1/3, \end{cases}$$

which is a rational (non-linear) function of x. Both reciprocals are presented in Fig. 1.8b. The dotted black line represents the α–cuts based reciprocal, and the dashed black line denotes the reciprocal obtained from (1.34). The results are very similar in that their supports and modal values coincide. The question is whether the approximation by a triangular fuzzy number does not imply a loss of information.

The division of fuzzy numbers was already defined using α–cuts based approach (1.22) and using the inversion of characteristic points (1.34). In the example below, the division of fuzzy numbers $\tilde{A}$ and $\tilde{B}$ is performed using different methods of multiplication of $\tilde{A}$ and the reciprocal of $\tilde{B}$ (Fig. 1.8).

Example 1.5 Let the following two triangular fuzzy numbers be given $\tilde{A} = (3, 5, 10)$, $\tilde{B} = (1, 2, 8)$ (see Fig. 1.9a). The division is performed here by first computing the reciprocal of $\tilde{B}$, and then by multiplying $\tilde{A}$ and $1/\tilde{B}$ using different multiplication formulae. The results are presented in Fig. 1.9b. The dashed grey line represents the quotient obtained using the combination of formulae (1.31) and (1.33). The dashed black line denotes the quotient obtained using the combination of formula (1.33) and, respectively, formulae (1.25) and (1.29) (the quotient obtained using the formula (1.26) coincides with this result). The grey solid line denotes the quotient obtained using the combination of formulae (1.33) and, respectively, formulae (1.27) and (1.29). The solid black line represents the quotient obtained using the α–cuts based approach. The membership function, in this case, is a rational function and has the following form:

$$\mu_{\tilde{C}}(x) = \begin{cases} \frac{8x-3}{2+6x}, & 0.375 \leqslant x < 2.5, \\ \frac{10-x}{5+x}, & 2.5 \leqslant x \leqslant 10. \end{cases} \tag{1.35}$$

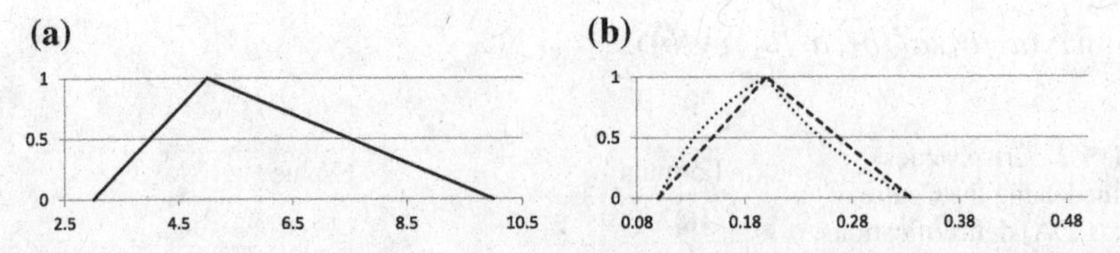

Fig. 1.8 The fuzzy number $\tilde{A} = (3, 5, 10)$ (**a**); reciprocal obtained using the inversion of characteristic points (*dashed black line*) and using. α–cuts (*dotted black line*) (**b**)

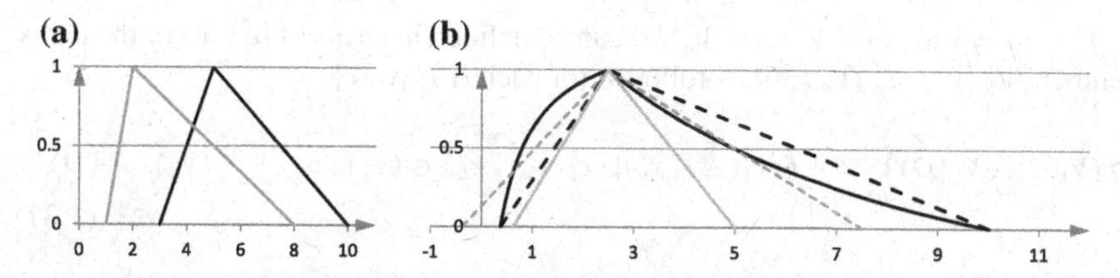

Fig. 1.9 Two exemplary triangular fuzzy numbers $\tilde{A} = (3, 5, 10)$ and $\tilde{B} = (1, 2, 8)$ (**a**); the results of computing $\tilde{A}/\tilde{B} = \tilde{A} * (1/\tilde{B})$ using different multiplication formulae (**b**)

Table 1.3 Crisp values obtained using the Centre of Area defuzzification method

Formula	Crisp number
(1.22)	6.35
(1.25), (1.29)	6.79
(1.27), (1.28)	5.21
(1.31)	5.71

As it was expected, the resulting fuzzy numbers differ significantly. The corresponding crisp numbers are summarised in Table 1.3. The difference between the smallest and the largest crisp value is greater than 1. An attempt to verify whether the choice of the respective approach to fuzzy arithmetic influences (or not) the final result will be made in Chap. 5.

1.3 Modelling Dependencies Between Fuzzy Numbers

Arithmetic operations on fuzzy numbers, presented in the previous section, implicitly assume that all combinations of implementation of fuzzy numbers involved are possible. In practical applications this assumption is rarely satisfied. For example, large implementations of crude oil prices usually involve large implementations of petrol prices. Combinations of small implementations of prices of one product and large prices of the other one are not likely to appear in reality. Ignoring this dependency may lead to significant overestimation of the exact result.

The dependency between variables (model parameters) can take different forms. When the dependency between variables $X_1, X_2, \ldots, X_n$ exists under the form of a domain $D \subseteq \Re^n$, which acts as the actual range of $(X_1, X_2, \ldots, X_n)$, and if, moreover, $X_1, X_2, \ldots, X_n$ have respective fuzzy ranges $\tilde{V}_1, \tilde{V}_2, \ldots, \tilde{V}_n$, then Zadeh's extension principle (1.15) must be adapted as follows [36]: for each $z \in \Re$,

$$\mu_{f(\tilde{V}_1, \tilde{V}_2, \ldots \tilde{V}_n | D)}(z) = \sup_{\substack{z = f(x_1, \ldots, x_n) \\ (x_1, \ldots, x_n) \in D}} \{\min(\mu_{\tilde{V}_1}(x_1), \ldots, \mu_{\tilde{V}_n}(x_n))\}. \tag{1.36}$$

The fuzzy set $f(\tilde{V}_1, \tilde{V}_2, \ldots, \tilde{V}_n|\tilde{D})$ can be defined in terms of α–cuts of the fuzzy numbers $\tilde{V}_1, \tilde{V}_2, \ldots, \tilde{V}_n$ [36] as follows: for each $\alpha \in [0, 1]$,

$$\left(f(\tilde{V}_1, \ldots, \tilde{V}_n|\tilde{D})\right)_\alpha = \{ f(x_1, \ldots, x_n)| (x_1, \ldots, x_n) \in (\tilde{V}_1)_\alpha \times \cdots \times (\tilde{V}_n)_\alpha \cap (\tilde{D})_\alpha\}. \quad (1.37)$$

Calculating a fuzzy quantity $f(\tilde{V}_1, \ldots, \tilde{V}_n|\tilde{D})$ using the Eq. (1.37) is difficult and very often involves nonlinear programming methods. Therefore, various methods for handling selected cases of interactivity between fuzzy variables have been developed. Dubois and Prade [27] solved the problem of arithmetic operations on fuzzy variables when the set $\tilde{D}$ is specified by a linear equation. Enea and Piazza [32] discussed the problem of interactive fuzzy numbers in the context of *Constrained Fuzzy AHP (CFAHP)*. Klir [33] presented the results of arithmetic operations on trapezoidal fuzzy numbers under requisite constraints. For example, he discusses the results of arithmetic operations on two fuzzy variables under the constraint that one variable is smaller than another and compares the results with the ones obtained without imposing this constraint. Kuchta [31] introduced the so-called constrained subtraction, which for $\tilde{A} = (a_1, a_2, a_3, a_4)$ and $\tilde{B} = (b_1, b_2, b_3, b_4)$ with $a_i - b_i \leqslant a_{i+1} - b_{i+1}$, for $i = 1, 2, 3$, is defined as follows:

$$\tilde{A} - \tilde{B} = (a_1 - b_1, a_2 - b_2, a_3 - b_3, a_4 - b_4)\,. \quad (1.38)$$

The constrained subtraction refers to the situations when large implementations of one variable are always accompanied by large implementations of another variable and small implementations of one variable by small implementations of another variable. However, when analysing economical facts, the link between the variables are not always that clear-cut.

1.3.1 Modelling Stochastic Dependencies

If the dependence between fuzzy variables is expressed in the form of the a correlation, then operations on those variables can be performed using the *fuzzy simulation*. The simulation of fuzzy systems was introduced by Liu and Iwamura [34]. It consists in executing random tests on a model with fuzzy parameters.

Given an univalued function f and a vector of correlated variables $\boldsymbol{X} = (X_1, X, \ldots, X_M) \in \Re^M$, for which values are limited by the respective elements of a fuzzy vector $\tilde{\boldsymbol{V}} = (\tilde{V}_1, \tilde{V}_2, \ldots, \tilde{V}_M)$, a computer fuzzy simulation presented in Algorithm 1 [35] can be used to specify the possibility distribution of $f(\tilde{\boldsymbol{V}})$ as well as its expected value $E(f(\tilde{\boldsymbol{V}}))$ and a standard semi-deviation $SDev(f(\tilde{\boldsymbol{V}}))$.

The elements of the vector $\boldsymbol{X} = (\mathrm{x}_1, \mathrm{x}_2, \ldots, \mathrm{x}_M)$ are correlated random numbers from the α–cuts of the respective fuzzy numbers $\tilde{V}_1, \tilde{V}_2, \ldots, \tilde{V}_M$. They are obtained in the following way. First, random values $z_1, z_2, \ldots, z_M$ are drawn from a uniform

Algorithm 1 Computing the value of a function for interactive fuzzy numbers

Input: Function f, vector of correlated fuzzy variables for which values are limited by the respective elements of the vector of fuzzy numbers $\tilde{V} = (\tilde{V}_1, \tilde{V}_2, \ldots, \tilde{V}_M)$

Output: Possibility distribution of $f(\tilde{V})$

1: Define $\alpha_0, \wp, \hat{n}, \dddot{n} = (1/\wp + 1)\hat{n}$
2: $\alpha = \alpha_0$
3: Define α-levels of the fuzzy numbers
4: $n = 1$
5: Generate a vector of correlated random values $(x_1, x_2, \ldots, x_M) \in \tilde{V}_1^\alpha \times \tilde{V}_2^\alpha \times \cdots \times \tilde{V}_M^\alpha$
6: $f_n = f(x_1, x_2, \ldots, x_m), \mu(f_n) = \min_{i=1,\ldots,M} \{\mu_{\tilde{V}_i}(x_i)\}$
7: $n = n + 1$
8: **If** $n \leqslant \hat{n}$ **Then Go To Step** 5
9: $\alpha = \alpha + \wp$
10: **If** $\alpha < 1$ **Then Go To Step** 3
11: Define possibility distribution $f(\tilde{V})$
11.1: $\alpha = 1, k = 1$
11.2: $Ls_k = \min_{1 \leqslant n \leqslant \dddot{n}} \{f_n \mid \mu(f_n) \geqslant \alpha\}, \mu(L_{s_k}) = \mu(f_n) | L_{s_k} = f_n$
$Rs_k = \max_{1 \leqslant n \leqslant \dddot{n}} \{f_n \mid \mu(f_n) \geqslant \alpha\}, \mu(R_{sk}) = \mu(f_n) | R_{sk} = f_n$
11.3: $\alpha = \alpha - \wp, k = k + 1$
11.4: **Repeat While** $\alpha > 0$
$Ls_k = \min_{1 \leqslant n \leqslant \dddot{n}} \{f_n | \mu(f_n) \geqslant \alpha \wedge f_n \leqslant Ls_1\}, \mu(L_{s_k}) = \mu(f_n) | L_{s_k} = f_n$
$Rs_k = \max_{1 \leqslant n \leqslant \dddot{n}} \{f_n | \mu(f_n) \geqslant \alpha \wedge f_n \leqslant Rs_1\}, \mu(R_{s_k}) = \mu(f_n) | R_{s_k} = f_n$
$\alpha = \alpha - \wp, k = k + 1$
Loop
12: Define credibility distribution function for $f(\tilde{V})$ on the basis of $L_{s_k}, R_{s_k}, \mu(L_{s_k})$ and $\mu(R_{s_k})$

distribution on the interval $[0, 1)$. Put $\boldsymbol{Z} = (z_1, z_2, \ldots, z_M)$ and let Ω be a correlation matrix describing dependencies between the considered fuzzy variables. Then, $Z = LZ$, where L is obtained from the Cholesky decomposition of the correlation matrix, i.e., $\Omega = LL^T$. In order to obtain the vector $\boldsymbol{X}$, the elements of the vector Z are mapped onto α–cuts of the respective fuzzy numbers. Once the vector $\boldsymbol{X}$ is obtained, the grade of membership of $z = f(\boldsymbol{X})$ is calculated as follows:

$$\mu_{f(\tilde{V})}(z) = \min_{i=1,2,\ldots,M} \{\mu_{\tilde{V}_i}(x_i)\}. \tag{1.39}$$

The method presented in Algorithm 1 allows to take into account the interactivity between the fuzzy parameters while computing the possibility distribution of $f(\tilde{V})$. The values Ls_k, Rs_k and $\mu(Ls_k)$, $\mu(Rs_k)$ specify the possibility distribution of the function f.

1.3.2 Interval Regression

An interval regression is identified when parameters of a regression equation are expressed in the form of bounded intervals [36–38]. Numerous practical problems [39] can be solved using interval and fuzzy regressions.

Until now, several methods were developed to estimate the parameters of the interval regression equation. The best known method uses a linear programming for this purpose [37, 40, 41]. However, this approach has many deficiencies. The major ones include [36–38]:

- It is often the case that some of the estimated regression parameters tend to be crisp; it even happens that the method produces only a few unexpectedly wide interval parameters, while others are crisp (the drawback called *unbalancedness*). This problem is generally considered to be the most serious and the most restrictive drawback that limits the usefulness of this method.
- The method might produce interval regression parameters with centres that only poorly fit the data with respect to traditional goodness-of-fit measures (such as R-squared). In literature, this problem is referred to as *non-centrality property*.
- The method is highly sensitive to outliers.

Many authors present solutions eliminating these deficiencies. Often, quadratic programming methods are combined with least squares methods [37, 40, 41]. Other methods of interval regression are based on Minkowski distance [42] or multi-criteria programming [43].

The abovementioned modified methods provide more balanced intervals representing the coefficients of interval regression equations, however, they require longer computation time. Moreover, the estimates of weight coefficients are made by experts, thus these methods become heuristic [39].

Another method for determining parameters of interval regression equations was developed by Hládik and Černy [39]. It seems to be very promising. Below is described a variant of this method which covers the case when input data (dependent and independent variables) are presented as determined figures (*crisp input-crisp output*). The proposed method is based on the sensitivity analysis of linear systems. It consists of two stages:

- the estimation of the centres of the interval parameters by using standard estimators,
- the estimation of the radii of the interval parameters.

Suppose that p observations are given:

$$\hat{y} = \begin{bmatrix} y_1 \\ \vdots \\ y_p \end{bmatrix}, X = \begin{bmatrix} x_{11} \dots x_{11} \\ \vdots \quad \vdots \\ x_{p1} \cdots x_{pn} \end{bmatrix}, \tag{1.40}$$

where X is an input matrix and $\hat{y}$ is an output vector. The problem is to determine the parameters of interval regression:

$$\hat{a} = \begin{bmatrix} a_1 \\ \vdots \\ a_n \end{bmatrix} \tag{1.41}$$

that comprise all possibilities determined by the model and data. More formally, $\hat{a}$ must be such that [39]:

$$y_j \in x_{j1}a_1 + x_{j2}a_2 + \cdots + x_{jn}a_n, \qquad \forall j = 1, \ldots, p. \tag{1.42}$$

It is natural to try to achieve several desirable properties of the interval parameters. They should be as tight as possible, they should be balanced and they should respect the central tendency. Easy handling with outliers is also an advantage.

Hladik and Černy [39] present interval regression coefficients in the form of intervals $\hat{a} = [\hat{b} - \hat{c}, \hat{b} + \hat{c}]$. Vector $\hat{b}$ is estimated by means of the least squares method. Width of intervals defined by the vector $\hat{c}$ is expressed in form: $\hat{c} = \delta\hat{c}^{\Delta}$, where $\hat{c}^{\Delta}$ is a non-negative vector of sensitivity coefficients and $\delta \geqslant 0$ is an unknown value. The introduction of sensitivity coefficients enable the width of intervals representing regression coefficients to be controlled.

Solution of the problem of estimation of interval regression parameters is reduced to finding the minimum value $\delta \geqslant 0$, such that for the vector $\hat{a} = [\hat{b} - \delta\hat{c}^{\Delta}, \hat{b} + \delta\hat{c}^{\Delta}]$ the following is met:

$$\forall j \in \{1, \ldots, p\}\ \exists \hat{a}' \in [\hat{b} - \delta\hat{c}^{\Delta},\ \hat{b} + \delta\hat{c}^{\Delta}] : y_j = X_{j*}\hat{a}'. \tag{1.43}$$

Hladik and Černy [39] present a simple formula for computing δ. Namely, when there exists $j \in \{1, \ldots, n\}$, such that $|X|_{j*}\hat{c}^{\Delta} = 0$ and simultaneously $y_j \neq X_{j*}\hat{b}$, then there doesn't exist δ, which would meet the interrelation (1.43). In any other case, δ may be calculated based on the following relation:

$$\delta^* = \max_{j : |X|_{j*} > 0} \frac{\left|y_j - X_{j*}\hat{b}\right|}{|X|_{j*}\hat{c}^{\Delta}}. \tag{1.44}$$

The value δ^* thus defined is a minimum value of δ. This relation constitutes an effective method for solving the problem of the interval regression. It eliminates the issue of significant imbalance of intervals representing particular regression coefficients and the problem of non-centric location of the estimated intervals.

Sensitivity coefficients are most often assumed as ${c_i}^{\Delta} = 1$ or ${c_i}^{\Delta} = |b_i|$ for $i = 1, \ldots, n$. The first case is one of the most natural choices since it forces the interval parameters to have a minimum sum of their radii. The second case is called relative tolerances. In this case, minimum value of δ (δ^*) gives the following information to a user: it is sufficient to perturb the regression parameters by no more than 100 % δ^* in order that all observations are satisfied. Hence, it is an alternative measure of goodness-of-fit.

1.4 Possibility and Credibility Theory

Possibility theory was first introduced by Zadeh [44] as an extension of theory of fuzzy sets and fuzzy logic. According to Dubois and Prade [45], each fuzzy set generates two functions: the measure of possibility *Pos* and the measure of necessity *Nec*. They are defined for every classic set $X \subseteq \Re$ as follows.

Definition 1.13 Let $X \subseteq \Re$ and let $\tilde{A} \in \mathcal{F}(\Re)$. The possibility measure Pos is a mapping $Pos : 2^X \to [0, 1]$ defined by

$$Pos(X) = \sup_{x \in X}\{\mu_{\tilde{A}}(x)\}, \quad (1.45)$$

where $\mu_{\tilde{A}}$ is the membership function of a fuzzy number $\tilde{A}$.

Definition 1.14 Let $X \subseteq \Re$ and let $\tilde{A} \in \mathcal{F}(\Re)$. The necessity measure Nec is a mapping $Nec : 2^X \to [0, 1]$ defined by:

$$Nec(X) = \inf_{x \notin X}\{1 - \mu_{\tilde{A}}(x)\}, \quad (1.46)$$

where $\mu_{\tilde{A}}$ is the membership function of a fuzzy number $\tilde{A}$.

These definitions are associated with the well-known interpretation of a fuzzy set given by Zadeh [13]. They assume that a fuzzy set $\tilde{A}$ is a fuzzy restriction of a certain variable X, which takes values in space $\Re$, and the only available information about this variable is that "X is $\tilde{A}$". If X is a variable that takes values in $\Re$ and $\tilde{A}$ is a fuzzy number characterised by a membership function $\mu_{\tilde{A}}$, then $\tilde{A}$ is a *fuzzy restriction* on X if $\tilde{A}$ acts as an elastic constraint on the values that may be assigned to X. In other words, the assignment of a value u, $u \in \Re$, to X has the form $X = u : \mu_{\tilde{A}}(u)$, where $X = u : \mu_{\tilde{A}}(u)$ is interpreted as the degree to which the constraint represented by $\tilde{A}$ is satisfied when u is assigned to X [44].

Based on the possibility distribution, it can be determined how possible is the event that the value of X belongs to a non-fuzzy set $A \subseteq \Re$, [1]:

$$\pi(X \in A) = \sup\{\pi_X(u) : u \in A\} = \sup\{\mu_{\tilde{A}}(u) : u \in A\} = \pi(A). \quad (1.47)$$

Such defined quantity does not have complementary characteristics, i.e., $\pi(X \in A)$ does not have to be equal to $1 - \pi(X \in A^c)$, where A^c is the absolute complement of A. As a remedy, Liu [46] introduced the concept of the *credibility measure*.

Definition 1.15 The set function $Cr : 2^{\Re} \rightarrow [0, 1]$ is called a credibility measure if it satisfies the following axioms:

1. $Cr(\Re) = 1$.
2. Cr is increasing, i.e., $Cr(A) \leqslant Cr(B)$ whenever $A \subset B$.
3. Cr is self-dual, i.e., $Cr(A) + Cr(A^c) = 1$ for any $A \in 2^{\Re}$.
4. $Cr\left(\bigcup_i A_i\right) \wedge 0.5 = \sup_i Cr(A_i)$for any A_i with $Cr(A_i) \leqslant 0.5$.

Definition 1.16 A fuzzy variable $\tilde{\xi}$ is defined a function from a credibility space $(\Re, 2^{\Re}, Cr)$ to the set of real numbers.

Definition 1.17 Let $\tilde{\xi}$ be a fuzzy variable and let A be a set of real numbers. The degree of credibility that the value of $\tilde{\xi}$ belongs to A can be defined as follows:

$$Cr(\tilde{\xi} \in A) = \frac{1}{2}(\pi(\tilde{\xi} \in A) + (1 - \pi(\tilde{\xi} \in A^c)). \tag{1.48}$$

Contrary to the possibility distribution, the degree of credibility has complementary characteristics [46]. Moreover, when the grade of credibility reaches a value of 1, there is a confidence that the fuzzy event will definitely occur. On the other hand, when the degree of possibility reaches a value of 1, such a confidence does not exist.

In addition to the credibility measure, Liu [46] has defined the concept of the *credibility distribution* $\Phi(x)$, which is for a fuzzy variable what is probability distribution for a random variable.

Definition 1.18 The credibility distribution $\Phi : \Re \rightarrow [0, 1]$ of a fuzzy variable $\tilde{\xi}$ is defined by $\Phi(x) = Cr\{u \in \Re | \tilde{\xi}(u) \leqslant x\}$.

The value $\Phi(x)$ determines the *grade of credibility* that a fuzzy variable $\tilde{\xi}$ will have a value equal to or less than x. If $\mu_{\tilde{\xi}}$ is a membership function of fuzzy variable $\tilde{\xi}$, then for each x in $\Re$, the credibility distribution function $\Phi(x)$ can be expressed by:

$$\Phi(x) = \frac{1}{2}(\sup_{y \leqslant x} \mu_{\tilde{\xi}}(y) + 1 - \sup_{y > x} \mu_{\tilde{\xi}}(y)). \tag{1.49}$$

1.4.1 Non-linear Programming Approach

The main problem addressed in this section is to determine possibility distribution of $f(\tilde{X})$. The vector $\tilde{X} = (X_1, X, \ldots, X_m)$ is a vector of the projected fuzzy variables about membership functions $\mu_1 \mu_2, \ldots, \mu_m$. Additionally, it is assumed that there may be defined subsets X^K of dependent variables X_i, $X^K = \left\{\tilde{X}_i, i \in K\right\}$, $K \in K_s$. In such case, K is the subset of dependent variable indices, and K_s is the set of indices of the selected subsets of dependent variables.

The problem of calculating the fuzzy number characterising $f(\tilde{X})$ can be solved by using the concept of α-levels of fuzzy numbers. Here one can take advantage of the Eq. (1.41). Upper bound (sup) and lower bound (inf) of an α-level of a fuzzy number $f(\tilde{X})$ may be determined by solving the following non-linear programming tasks:

when searching for maximum, find:

$$f(x_1, x_2, \ldots, x_m) \rightarrow \max \tag{1.50}$$

when searching for minimum, find:

$$f(x_1, x_2, \ldots, x_m) \rightarrow \min \tag{1.51}$$

subject to the following constraints:

$$\inf(\tilde{V}_i)_\alpha \leqslant x_i \leqslant \sup(\tilde{V}_i)_\alpha \quad \text{for } i = 1, 2 \ldots, m, \tag{1.52}$$

$$x_i \geqslant \inf(a_1^{iz})x_z + \inf(a_2^{iz}) \quad \text{for } i \in K, z \in K;\ i \neq z;\ K \in K_s, \tag{1.53}$$

$$x_i \leqslant \sup(a_1^{iz})x_z + \sup(a_2^{iz}) \quad \text{for } i \in K, z \in K;\ i \neq z;\ K \in K_s, \tag{1.54}$$

The values a_1^{iz}, a_2^{iz} are the coefficients of *interval regression* equations. They determine a relation between variables $\hat{X}_i$ and $\hat{X}_z$ based on historical data.

References

1. Zadeh, L.A. 1965. Fuzzy sets. *Information and Control* 8(3): 338–353.
2. Wang, L.-X. 1997. *A course in fuzzy systems and control*. Prentice Hall.
3. Zadeh, L.A. 1971. Similarity relations and fuzzy orderings. *Information Sciences* 3(2): 177–200.
4. Heilpern, S. 1995. Comparison of fuzzy numbers in decision making. *Tatra Mountains Mathematical Publications* 6: 47–53.
5. Zimmermann, H.-J. 1996. *Fuzzy Set Theory and its Applications*, 3rd ed. MA: Kluwer Academic Publishers Norwell.
6. Pedrycz, A. 2006. An optimization of ac-cuts of fuzzy sets through particle swarm optimization. *Fuzzy information processing society. NAFIPS 2006. Annual meeting of the North American, IEEE*, 57–62.
7. Carlsson, C., and R. Fullèr. 2005. On additions of interactive fuzzy numbers. *Acta Polytechnica Hungarica* 2: 59–73.
8. Höhle, U. 1978. Probabilistic uniformization of fuzzy topologies. *Fuzzy Sets and Systems* 1.
9. Alsina, C., E. Trillas, and L. Valverde. 1983. On some logical connectives for fuzzy set theory. *Journal of Mathematical Analysis & Applications* 93: 15–26.
10. Klement, E.P. 1982. A theory of fuzzy measures: A survey. In *Fuzzy information and decision processes*, ed. M. Gupta, and E. Sanchez, 59–66. Amsterdam: North Holland.
11. Dubois, D., and H. Prade. 1981. Triangular norms for fuzzy sets. In *International symposium of fuzzy sets*, 39–68, Linz.

12. Bonissone, P.P. 1985. Selecting uncertainty calculi and granularity: An experiment in trading off precision and complexity. In *Proceedings of the first Workshop on Uncertainty in Artificial Intelligence*, 57–66, Los Angeles.
13. Zadeh, L.A. 1975. In *Fuzzy sets and their applications to cognitive and decision processes*, ed. Zadeh, L.A., K.-S. Fu, K. Tanaka, and M. Shimura., Calculus of fuzzy restrictions.
14. Zadeh, L.A. 1973. Outline of a new approach to the analysis of complex systems and decision processes. *IEEE Transactions on Systems, Man, and Cybernetics* 3: 28–44.
15. Schweizer, B., and A. Sklar. 1960. Statistical metric spaces. *Pacific Journal of Mathematics* 10: 313–334.
16. Dubois, D., and H. Prade. 1980. *Fuzzy sets and systems: Theory and applications*, 4th ed., Tom 144: Mathematics in science and engineering New York: Academic Press.
17. Filev, D., and R. Yager. 1997. Operations on fuzzy numbers via fuzzy reasoning. *Fuzzy Sets and Systems* 91(2): 137–142.
18. Kreinovich, V., W. Pedrycz, and H.T. Nguyen. 2001. How to make sure that $/approx 100$ in fuzzy arithmetic: solution and its (inevitable) drawbacks. In *Proceedings of the joint 9th world congress of the international fuzzy systems association and 20th international conference of the North American fuzzy information processing society IFSA/NAFIPS*, 1653–1658, Vancouver, Canada, July 25–28.
19. Fodor, J., and B. Bede. 2006. Recent advances in fuzzy arithmetics. *International Journal of Computers, Communications and Control* 1(5): 199–207.
20. Tsao, C.T. 2002. An interval arithmetic approach for fuzzy multiple criteria group decision making. *Asia Pacific Management Review* 7(1): 25–40.
21. Gao, S., Z. Zhang, and C. Cao. 2009. Multiplication operation on fuzzy numbers. *Journal of Software* 4(4): 331–338.
22. Dutta, P., H. Boruah, and T. Ali. 2011. Fuzzy arithmetic with and without using $/alpha$-cut method: A comparative study. *International Journal of Latest Trends in Computing* 2(1): 99–107.
23. Bansal, A. 2011. Trapezoidal fuzzy numbers (a, b, c, d): Arithmetic behavior. *International Journal of Physical and Mathematical Sciences* 2(1): 39–44.
24. Banerjee, S., and T.K. Roy. 2012. Arithmetic operations on generalized trapezoidal fuzzy number and its applications. *Turkish Journal of Fuzzy Systems* 3(1): 16–44.
25. Vahidi, J., and S. Rezvani. 2013. Arithmetic operations on trapezoidal fuzzy numbers. *Journal Nonlinear Analysis and Application* 2013: 1–8. www.ispacs.com/jnaa. Accessed 20 Feb 2015.
26. Kolesárová, A. A note on shape preserving t-norm-based multiplication of fuzzy numbers. http://www.polytech.univ-savoie.fr/fileadmin/polytech_autres_sites/sites.
27. Dubois, D., and H. Prade. 1988. In *Analysis of fuzzy information*, 2nd ed, ed. Bezdek, J.C., 3–39., Fuzzy numbers: An overview Boca Raton: CRC Press.
28. Dubois, D., and H. Prade. 1982. Towards fuzzy differential calculus part 1: Integration of fuzzy mappings. *Fuzzy Sets and Systems* 8(1): 1–17.
29. Moore, R.E. 1966. *Interval analysis.*, Prentice-Hall, Series in automatic computation New Jersey: Prentice-Hall.
30. Hong, D.H. 2001. Shape preserving multiplications of fuzzy numbers. *Fuzzy Sets and Systems* 123(1): 81–84.
31. Kuchta, D. 2001. *Miękka matematyka w zarządzaniu: Zastosowanie liczb przedziałowych i rozmytych w rachunkowości zarządczej*. Wrocław: Oficyna Wydawnicza Politechniki Wrocławskiej.
32. Enea, M., and T. Piazza. 2004. Project selection by constrained fuzzy ahp. *Fuzzy Optimization and Decision Making* 3(1): 39–62.
33. Klir, G.J. 1997. Project selection by constrained fuzzy ahp. *Fuzzy Sets and Systems - Special issue: Fuzzy Arithmetic* 91(2): 165–175.
34. Liu, B., and K. Iwamura. 2001. Fuzzy programming with fuzzy decisions and fuzzy simulation-based genetic algorithm. *Fuzzy Sets and Systems* 122(2): 253–262.
35. Rębiasz, B. 2013. Selection of efficient portfoliosprobabilistic and fuzzy approach, comparative study. *Computers & Industrial Engineering* 64(4): 1019–1032.

36. Tanaka K., S. Uejima, and K. Asai. 1982. Linear regression analysis with fuzzy model. *IEEE Transactions on Systems, Man, and Cybernetics, SMC-12*, 6:903–907.
37. Tanaka, H. 1987. Fuzzy data analysis by possibilistic linear models. *Fuzzy Sets and Systems* 24: 363–375.
38. Tanaka, H., and H. Lee. 1998. Interval regression analysis by quadratic programming approach. *IEEE Transactions on Fuzzy Systems* 6: 473–481.
39. Hladík, M., and M. Černý. 2012. Interval regression by tolerance analysis approach. *Fuzzy Sets and Systems* 193(1): 85–107.
40. Lee, H., and H. Tanaka. 1999. Lower and upper approximation models in interval regression using regression quantile techniques. *European Journal of Operational Research* 116: 653–666.
41. Tanaka, H., and J. Watada. 1988. Possibilistic linear systems and their application to the linear regression model. *Fuzzy Sets and Systems* 27: 275–289.
42. Inuiguchi, M., H. Fujita, and T. Tanino. 2002. Robust interval regression analysis based on minkowski difference. *Proceedings of the 41st SICE annual conference SICE 2002*, vol. 4, 2346–2351, Japan: Osaka.
43. Tran, L., and L. Duckstein. 2002. Multiobjectine fuzzy regression with central tendency and possibilistic properties. *Fuzzy Sets and Systems* 130: 21–31.
44. Zadeh, L.A. 1978. Fuzzy sets as a basis for a theory of possibility. *Fuzzy Sets and Systems* 1(1): 3–28.
45. Dubois, D., and H. Prade. 1983. Ranking of fuzzy numbers in the setting of possibility theory. *Information Sciences* 30(3): 183–224.
46. Liu, B. 2006. A survey of credibility theory. *Fuzzy Optimization and Decision Making* 5(4): 387–408.

Chapter 2
Ordering of Fuzzy Numbers

Abstract This chapter describes different methods for comparing and ordering fuzzy numbers. Theoretically, fuzzy numbers can only be partially ordered, and hence cannot be compared. However, in practical applications, such as decision making, scheduling, market analysis or optimisation with fuzzy uncertainties, the comparison of fuzzy numbers becomes crucial.

Theoretically, fuzzy numbers can only be partially ordered, and hence cannot be compared. However, when they are used in practical applications, e.g., when a decision must be made among alternatives or an optimal value of an objective function must be found, the comparison of fuzzy numbers becomes crucial.

There are numerous approaches to the ordering relation between fuzzy numbers [1–6] qualitative, quantitative and based on α–cuts. Jain [7] and Dubois and Prade [4] were the first who considered this problem. Some methods to rank fuzzy numbers were reviewed by Bortolan and Degani [8]. Detyniecki and Yager [3] proposed the α-weighted valuations of fuzzy numbers. Hong and Kim [9] proposed an easy way to compute the min and max operation for fuzzy numbers. Asady and Zendehnam [10] proposed the ranking fuzzy numbers by distance minimisation method. Comparison of various ranking methods for fuzzy numbers with the possibility of ranking the crisp numbers was described by Thorani et al. [11]. The problem of comparing of fuzzy numbers was also considered by Allahviranloo et al. [12]. They proposed a method based on the centroid point of a fuzzy number and its area. Sevastjanov and Róg [13] developed a probability-based comparison of fuzzy numbers. The probabilistic approach was also considered in [14]. The large number of fuzzy ordering methods can be justified by the fact that different methods can be useful for different purposes. For example, problems involving ranking, prioritising or choosing between large number of alternatives will benefit from methods that assign to fuzzy numbers crisp values thus reducing the fuzzy ordering problem to ordering of real numbers.

An overview of selected approaches to ordering (ranking) of fuzzy numbers is presented below. The presented approaches can be generally divided into two groups. The first group consists of methods which enable two fuzzy numbers to be compared. Included in this group are such methods as probabilistic approach, centroid point approach or radius of gyration approach. To order a set of fuzzy numbers using

© Springer International Publishing Switzerland 2015

I. Skalna et al., *Advances in Fuzzy Decision Making*,
Studies in Fuzziness and Soft Computing 333,
DOI 10.1007/978-3-319-26494-3_2

these methods, some dedicated procedures are required. The second group consists of methods, which assign to a fuzzy number a crisp value. These are methods such as Yager ranking index based approach, defuzzification approach or weighted average. The methods from the second group can be directly used to order a set of fuzzy numbers, by employing one of the several methods for ordering (sorting) real numbers. All the above mentioned methods are compared using an example of ordering four triangular fuzzy numbers.

2.1 Probabilistic Approach

The *probabilistic* (also known as *probability degree-based* or *probability-based*) approach to ordering fuzzy numbers is based on the α–cuts representation of fuzzy numbers. The α–cuts based orderings are so attractive, because they can be used regardless the type of the membership function. Moreover, each α-level is an interval, so the powerful tools of interval arithmetic [15] can be employed to solve the problem of fuzzy ordering [13].

Let $\boldsymbol{a} = [a_1, a_2]$ and $\boldsymbol{b} = [b_1, b_2]$ be two closed and compact intervals. The possibility degree-based ranking method which is shown in Table 2.1 was proposed by Jiang et al. [16]. The non-overlapping cases are omitted as they are obvious.

A similar, but slightly extended approach to ordering of intervals was proposed in [13]. Let the real values $a \in \boldsymbol{a}$ and $b \in \boldsymbol{b}$ be given. They can be considered as two independent uniform random variables. If $\boldsymbol{a}$ and $\boldsymbol{b}$ overlaps, then some disjoint subintervals can be distinguished. The fall of random variables a and b in the subintervals $[a_1, b_1]$, $[b_1, a_2]$, $[a_2, b_2]$ may be treated as a set of independent random events.

Let the events $H_k : a \in \boldsymbol{a}_i,\ b \in \boldsymbol{b}_j$ be defined for $k = 1, \ldots, n$, where $\boldsymbol{a}_i$ and $\boldsymbol{b}_j$ are certain subintervals of intervals $\boldsymbol{a}$ and $\boldsymbol{b}$ in accordance with $\boldsymbol{a} = \bigcup_i \boldsymbol{a}_i$ and $\boldsymbol{b} = \bigcup_i \boldsymbol{b}_i$ ($n = 4$ for the case depicted in Fig. 2.1) [13]. Let $P(H_k)$ be the probability of event H_k, and $P(\boldsymbol{b} > \boldsymbol{a}|H_k)$ be the conditional probability of $\boldsymbol{b} > \boldsymbol{a}$ given H_k. Hence, the composite probability may be expressed as follows:

$$P(\boldsymbol{b} > \boldsymbol{a}) = \sum_{k=1}^{n} P(H_k)P(\boldsymbol{b} > \boldsymbol{a}|H_k). \tag{2.1}$$

Table 2.1 Cases of interval comparison proposed by Jiang [16]

Case	$P(\boldsymbol{b} \leqslant \boldsymbol{a})$
1. $b_1 \geqslant a_1 \wedge b_2 \geqslant a_2 \wedge b_1 \leqslant a_2$	$\frac{a_1-b_1}{b_2-b_1} \cdot \frac{a_2-a_1}{b_2-b_1}$
2. $a_1 \geqslant b_1 \wedge a_2 \leqslant b_2$	$\frac{a_1-b_1}{b_2-b_1} \cdot \frac{1}{2}\frac{a_2-a_1}{b_2-b_1}$
3. $a_1 \geqslant b_1 \wedge a_2 \geqslant b_2 \wedge a_1 \leqslant b_2$	$\frac{a_1-b_1}{b_2-b_1} + \frac{b_2-a_1}{b_2-b_1} \cdot \frac{a_2-b_2}{a_2-a_1} + \frac{1}{2}\frac{b_2-a_1}{b_2-b_1} \cdot \frac{b_1-a_1}{a_2-a_1}$
4. $b_1 \geqslant a_1 \wedge b_2 \leqslant a_2$	$\frac{a_2-b_2}{a_2-a_1} + \frac{1}{2}\frac{b_2-b_1}{a_2-a_1}$

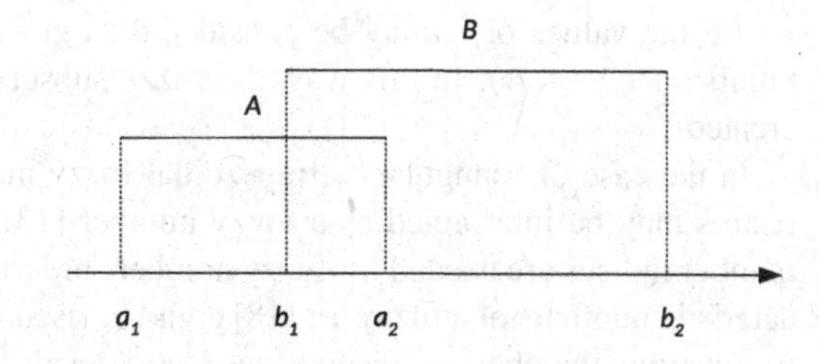

Fig. 2.1 Example of overlapping intervals

The resulting formula for the case of overlapping intervals is as follows:

$$P(\boldsymbol{b} > \boldsymbol{a}) = 1 - \frac{1}{2}\frac{(a_2-b_1)^2}{(a_2-a_1)(b_2-b_1)}. \tag{2.2}$$

The results obtained in [13] for all possible interval overlapping are shown in Table 2.2.

The above possibility degree-based methods have the following features:

1. $0 \leqslant P(\boldsymbol{a} \leqslant \boldsymbol{b}) \leqslant 1$.
2. If $P(\boldsymbol{b} \leqslant \boldsymbol{a}) \leqslant \alpha$, then $P(\boldsymbol{a} \leqslant \boldsymbol{b}) \leqslant 1 - \alpha$.
3. If $P(\boldsymbol{b} \leqslant \boldsymbol{a}) = P(\boldsymbol{a} \leqslant \boldsymbol{b})$, then $\boldsymbol{a} \equiv \boldsymbol{b}$.

It follows from 2 and 3 that if $\boldsymbol{a} \equiv \boldsymbol{b}$, then $P(\boldsymbol{b} \leqslant \boldsymbol{a}) = P(\boldsymbol{a} \leqslant \boldsymbol{b}) = 0.5$.

This approach can be treated as a framework for elaboration of constructive methods of interval comparison in various special situations. Some aspects of the interval comparison and ordering group of intervals, based on this approach, is presented, e.g., in [17].

Now, let $\tilde{A}$ and $\tilde{B}$ be some arbitrary fuzzy numbers, and let $\tilde{A}^\alpha = \{x \mid \mu_A(x) \geqslant \alpha\}$ and $\tilde{B}^\alpha = \{x \mid \mu_B(x) \geqslant \alpha\}$ be their respective α–cuts. Since $\tilde{A}^\alpha$ and $\tilde{B}^\alpha$ are intervals, probability $P^\alpha(\tilde{B}^\alpha > \tilde{A}^\alpha)$ for each pair $\tilde{A}^\alpha$ and $\tilde{B}^\alpha$ can be calculated in the way described in the previous section. The set of probabilities P^α, $\alpha \in (0, 1]$, may be treated as the support of the fuzzy subset [13]:

$$P(\tilde{A} > \tilde{B}) = \{\alpha \mid P^\alpha(\tilde{B}^\alpha > \tilde{A}^\alpha)\}, \tag{2.3}$$

Table 2.2 Typical cases of interval comparison [13]

Case	$P(a > b)$	$P(a = b)$
1. $a_1 > b_1 \wedge a_1 < b_2 \wedge a_1 = a_2$	$\frac{b_2-a_1}{b_2-b_1}$	0
2. $b_1 > a_1 \wedge b_1 < a_2 \wedge b_1 = b_2$	$\frac{b_1-a_1}{a_2-a_1}$	0
3. $b_1 \geqslant a_1 \wedge b_2 \leqslant a_2$	$\frac{b_1-a_1}{a_2-a_1} + \frac{1}{2}\frac{a_2-a_1}{b_2-b_1}$	$\frac{b_2-b_1}{a_2-a_1}$
4. $a_1 \geqslant b_1 \wedge a_2 \leqslant b_2$	$\frac{b_2-a_2}{b_2-b_1} + \frac{1}{2}\frac{a_2-a_1}{b_2-b_1}$	$\frac{a_2-a_1}{b_2-b_1}$
5. $b_1 \geqslant a_1 \wedge b_2 \geqslant a_2 \wedge b_1 \leqslant a_2$	$1 - \frac{1}{2}\frac{(a_2-b_1)^2}{(a_2-a_1)(b_2-b_1)}$	$\frac{(a_2-b_1)^2}{(a_2-a_1)(b_2-b_1)}$
6. $a_1 \geqslant b_1 \wedge a_2 \geqslant b_2 \wedge a_1 \leqslant b_2$	$1 - \frac{1}{2}\frac{(b_2-a_1)^2}{(a_2-a_1)(b_2-b_1)}$	$\frac{(b_2-a_1)^2}{(a_2-a_1)(b_2-b_1)}$

where the values of α may be considered as grades of membership of the fuzzy number $P(\tilde{A} > \tilde{B})$. In this way, the fuzzy subset $P(\tilde{A} = \tilde{B})$ may also be easily created.

In the case of triangular or trapezoidal fuzzy number comparison, the obtained results may be interpreted as a fuzzy number [13]. Nevertheless, in practice, real number indices are needed for fuzzy numbers ordering. For this purpose, some characteristic numbers of a fuzzy set [18] could be used. It seems, however, more natural to substitute the obtained discrete set I of α-levels with a real number:

$$\overline{P}(\tilde{B} > \tilde{A}) = \sum_{\alpha \in I} \alpha P^{\alpha}(\tilde{B}^{\alpha} > \tilde{A}^{\alpha}) / \sum_{\alpha \in I} \alpha. \tag{2.4}$$

The equation (2.4) emphasises that the contribution of the α-level to the overall probability estimation is increasing with an increase in its number. Of course, as proposed in [3], the set of complementary parametrised functions of α can be applied in the equation (2.4) instead of α.

Example 2.1 Let the following four triangular fuzzy numbers, depicted in Fig. 2.2, be given [19]:

$$\tilde{A}_1 = (0.12, 0.19, 0.29),\ \tilde{A}_2 = (0.22, 0.32, 0.48),$$

$$\tilde{A}_3 = (0.11, 0.15, 0.23),\ \tilde{A}_4 = (0.21, 0.33, 0.49).$$

The results of pairwise comparison of these numbers using the probability degree-based approach are presented in Table 2.3. This gives the following order: $\tilde{A}_3 < \tilde{A}_1 < \tilde{A}_2 < \tilde{A}_4$.

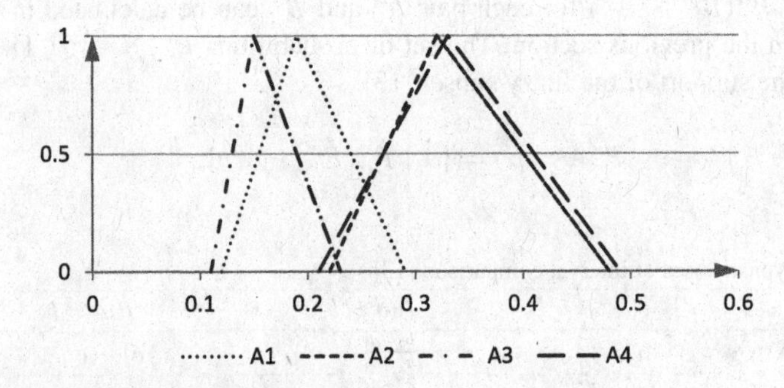

Fig. 2.2 Exemplary triangular fuzzy numbers

Table 2.3 The results of comparison of triangular fuzzy numbers $\tilde{A}_1, \tilde{A}_2, \tilde{A}_3, \tilde{A}_4$

Pair	$\overline{P}$	Comparison
$\tilde{A}_1, \tilde{A}_2$	0.999	$\tilde{A}_1 < \tilde{A}_2$
$\tilde{A}_1, \tilde{A}_3$	0.059	$\tilde{A}_1 > \tilde{A}_3$
$\tilde{A}_1, \tilde{A}_4$	0.998	$\tilde{A}_1 < \tilde{A}_4$
$\tilde{A}_2, \tilde{A}_3$	3.00E-08	$\tilde{A}_2 > \tilde{A}_3$
$\tilde{A}_2, \tilde{A}_4$	0.6823	$\tilde{A}_2 < \tilde{A}_4$
$\tilde{A}_3, \tilde{A}_4$	1	$\tilde{A}_3 < \tilde{A}_4$

2.2 Defuzzification Approach

Fuzzy numbers can also be ranked using the *defuzzification* methods. A defuzzification is the process of producing a real (crisp) value corresponding to a fuzzy number. In order to rank fuzzy numbers using the defuzzification approach, the fuzzy numbers are first defuzzified and then, the obtained crisp numbers are ordered using the order relation of real numbers. There are several defuzzification methods, among them:

- Centre of area (*COA*) or centre of gravity (*COG*):

$$COA(\tilde{A}) = COG(\tilde{A}) = \frac{\int_{x_{\min}}^{x_{\max}} x\mu_{\tilde{A}}(x)dx}{\int_{x_{\min}}^{x_{\max}} \mu_{\tilde{A}}(x)dx}.$$

- First of maxima (*FOM*):

$$FOM(\tilde{A}) = \min \ker(\tilde{A}).$$

- Middle of maxima (*MOM*):

$$MOM(\tilde{A}) = \frac{\min \ker(\tilde{A}) + \max \ker(\tilde{A})}{2}.$$

- Last of maxima (*LOM*):

$$LOM(\tilde{A}) = \max \ker(\tilde{A}).$$

In the case of triangular fuzzy numbers of the form $\tilde{A} = (a, b, c)$:

$$FOM(\tilde{A}) = MOM(\tilde{A}) = LOM(\tilde{A}) = c.$$

The results of ordering of fuzzy numbers from Example 2.1 using the abovementioned defuzzification methods are the same as the result from Example 2.1. The values of *FOM*, *MOM* and *LOM* are obvious, whereas $COG(\tilde{A}_1) = 0.2$, $COG(\tilde{A}_2) = 0.34$, $COG(\tilde{A}_3) = 0.1633$, $COG(\tilde{A}_4) = 0.3452$.

2.3 Centroid-Point Approach

A fuzzy number $\tilde{A}$ can be identified with an ordered pair of continuous real functions defined on the interval [0, 1], i.e., $\tilde{A} = (f_{\tilde{A}}, g_{\tilde{A}})$, where $f_{\tilde{A}}, g_{\tilde{A}} : [0,1] \to \Re$ are continuous functions. Functions $f_{\tilde{A}}$ and $g_{\tilde{A}}$ are called, respectively, the *up* and *down*-parts of a fuzzy number $\tilde{A}$.

The continuity of the functions f and g implies that their images are bounded intervals (see Fig. 2.3a) denoted, respectively, as UP and $DOWN$. If, additionally, the $f_{\tilde{A}}$ and $g_{\tilde{A}}$ functions are monotone, and thus invertible, the following membership can be defined:

$$\mu_{\tilde{A}}(x) = \begin{cases} f_{\tilde{A}}^{-1}(x), & x \in [f_{\tilde{A}}(0), f_{\tilde{A}}(1)] = [l_{\tilde{A}}, 1_{\tilde{A}}^{-}], \\ g_{\tilde{A}}^{-1}(x), & x \in [g_{\tilde{A}}(1), g_{\tilde{A}}(0)] = [1_{\tilde{A}}^{+}, p_{\tilde{A}}], \\ 1, & x = [1_{\tilde{A}}^{-}, 1_{\tilde{A}}^{+}], \end{cases} \tag{2.5}$$

if $f_{\tilde{A}}$ is increasing and $g_{\tilde{A}}$ is decreasing, and $f_{\tilde{A}} \leqslant g_{\tilde{A}}$ for all $y \in [0, 1]$. The obtained membership function $\mu_{\tilde{A}}(x), x \in \Re$ represents a mathematical object which resembles a convex fuzzy number in the classical sense.

Definition 2.1 Let $\tilde{A} = (a, b, c, d)$. Then, the centroid (centre of gravity) point of $\tilde{A}$ is obtained as follows [20]:

$$COGP(\tilde{A}) = (\overline{x}_0(\tilde{A}), \overline{y}_0(\tilde{A})), \tag{2.6}$$

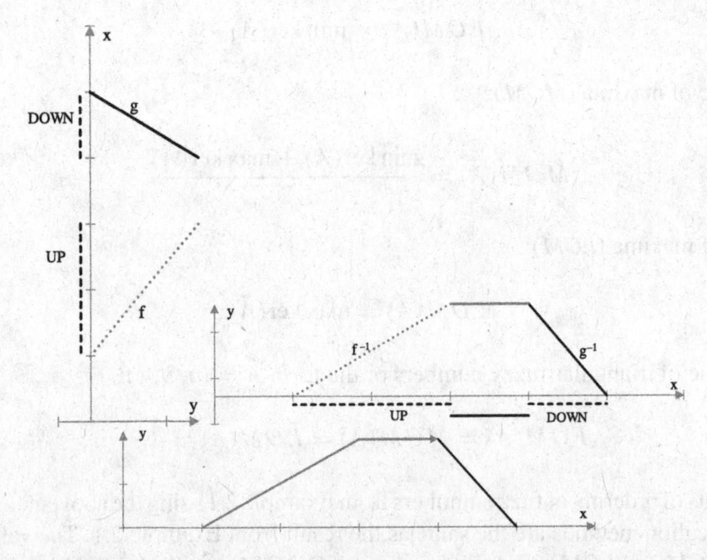

Fig. 2.3 An ordered fuzzy number (**a**), an ordered fuzzy number presented as a fuzzy number in a classical sense (**b**), and a simplified mark denoting the order of inverted functions (**c**)

where

$$\begin{cases} \overline{x}_0(\tilde{A}) = \dfrac{\int_a^b x f_{\tilde{A}}^{-1}(x)dx + \int_b^c x dx + \int_c^d x g_{\tilde{A}}^{-1}(x)dx}{\int_a^b f_{\tilde{A}}^{-1}(x)dx + \int_b^c dx + \int_c^d g_{\tilde{A}}^{-1}(x)dx}, \\ \overline{y}_0(\tilde{A}) = \dfrac{\int_0^1 y f_{\tilde{A}}(y)dy + \int_0^1 y g_{\tilde{A}}(y)dy}{\int_0^1 f_{\tilde{A}}(y)dy + \int_0^1 g_{\tilde{A}}(y)dy}. \end{cases} \tag{2.7}$$

In the case of trapezoidal fuzzy numbers, the above formula takes the form:

$$\begin{cases} \overline{x}_0(\tilde{A}) = \frac{1}{3}\left[a+b+c+d - \frac{cd-ab}{(c+d)-(a+b)}\right], \\ \overline{y}_0(\tilde{A}) = \frac{1}{3}\left[1 + \frac{c-b}{(c+d)-(a+b)}\right]. \end{cases} \tag{2.8}$$

Based on a centroid point, two fuzzy numbers $\tilde{A}$ and $\tilde{B}$ are compared using the following rules [21]:

$$\begin{aligned} &\text{If } \overline{x}_0(\tilde{A}) > \overline{x}_0(\tilde{B}), \text{ Then } \tilde{A} > \tilde{B}. \\ &\text{If } \overline{x}_0(\tilde{A}) < \overline{x}_0(\tilde{B}), \text{ Then } \tilde{A} < \tilde{B}. \\ &\text{If } \overline{x}_0(\tilde{A}) = \overline{x}_0(\tilde{B}), \text{ Then} \\ &\quad \text{If } \overline{y}_0(\tilde{A}) > \overline{y}_0(\tilde{B}), \text{ Then } \tilde{A} > \tilde{B}. \\ &\quad \text{Else If } \overline{y}_0(\tilde{A}) < \overline{y}_0(\tilde{B}), \text{ Then } \tilde{A} < \tilde{B}. \\ &\quad \text{Else} \tilde{A} = \tilde{B}. \end{aligned} \tag{2.9}$$

For the fuzzy numbers from Example 2.1 the following centroid points were obtained (Table 2.4):

This gives the ordering: $\tilde{A}_3 < \tilde{A}_1 < \tilde{A}_2 < \tilde{A}_4$. The method is rather simple, but it requires a pairwise comparison of fuzzy numbers to be ordered.

Table 2.4 The considered fuzzy numbers and their centroid points

Fuzzy number	$\overline{x}_0$	$\overline{y}_0$
$\tilde{A}_1 = (0.12, 0.19, 0.29)$	0.343	0.333
$\tilde{A}_2 = (0.22, 0.32, 0.48)$	0.34	0.333
$\tilde{A}_3 = (0.11, 0.15, 0.23)$	0.2	0.333
$\tilde{A}_4 = (0.21, 0.33, 0.49)$	0.163	0.333

Table 2.5 Yager index for the fuzzy numbers from Example 2.1

Fuzzy number	Yager index
$\tilde{A}_1 = (0.12, 0.19, 0.29)$	0.1317
$\tilde{A}_2 = (0.22, 0.32, 0.48)$	0.2233
$\tilde{A}_3 = (0.11, 0.15, 0.23)$	0.1067
$\tilde{A}_4 = (0.21, 0.33, 0.49)$	0.2267

2.4 Yager Ranking Index Approach

In [22], Yager proposed the following index to ordering fuzzy numbers:

$$Y(\tilde{A}) = \frac{1}{2}\int_{0}^{1} (f_{\tilde{A}}(y) + g_{\tilde{A}}(y))dy.$$

For example, given two triangular fuzzy numbers $\tilde{A}_1 = (35, 50, 61)$ and $\tilde{A}_2 = (30, 41, 49)$ the Yager's values are $Y(\tilde{A}_1) = 49$ and $Y(\tilde{A}_2) = 40.25$. Thus, $\tilde{A}_2$ is smaller than $\tilde{A}_1$ in the context of Yager index.

For the fuzzy numbers from Example 2.1, the Yager index takes the values presented in Table 2.5. They yield exactly the same order as the one obtained using the possibilistic approach.

2.5 Degree of Possibility Approach

The ordering of fuzzy numbers using priority approach is based on the research presented in [19].

Definition 2.2 Given two convex fuzzy numbers $\tilde{A}$ and $\tilde{B}$ the degree of possibility of $\tilde{A} > \tilde{B}$ is defined as

$$V(\tilde{A} > \tilde{B}) = \sup_{x \geq y}\{min\{\mu_{\tilde{A}}(x), \mu_{\tilde{B}}(y)\}\}. \tag{2.10}$$

Thus, if there exists a pair (x, y) such that $x \geqslant y$ and $\mu_{\tilde{A}}(x) = \mu_{\tilde{B}}(y) = 1$, then the degree of possibility $V(\tilde{A} > \tilde{B}) = 1$. Since $\tilde{A}$ and $\tilde{B}$ are convex fuzzy numbers, the following holds [19]:

$$\begin{aligned} &V(\tilde{A} > \tilde{B}) = 1 \text{ iff } \sup \ker(\tilde{A}) \geqslant \inf \ker(\tilde{B}), \\ &V(\tilde{B} \geqslant \tilde{A}) = \text{hgt}(\tilde{A} \cap \tilde{B}) = \mu_{\tilde{A}}(d), \end{aligned} \tag{2.11}$$

where d is the x-coordinate of the highest intersection point between $\mu_{\tilde{A}}$ and $\mu_{\tilde{B}}$. When $\tilde{A} = (a_1, a_2, a_3)$ and $\tilde{b} = (b_1, b_2, b_3)$, then

$$\text{hgt}(\tilde{A} \cap \tilde{B}) = \frac{a_1 - b_3}{(b_2 - b_3) - (a_2 - a_1)}. \quad (2.12)$$

To compare $\tilde{A}$ and $\tilde{B}$ both values $V(\tilde{A} \geqslant \tilde{B})$ and $V(\tilde{B} \geqslant \tilde{A})$ are needed.

Now, the degree of possibility for a convex fuzzy number $\tilde{A}$ to be greater than k convex fuzzy numbers $\tilde{A}_i$ $(i = 1, \ldots, k)$ is given by

$$\begin{aligned} V(\tilde{A} \geq \tilde{A}_1, \ldots, \tilde{A}_k) &= V[(\tilde{A} \geq \tilde{A}_1) \wedge (\tilde{A} \geq \tilde{A}_2) \wedge \cdots \wedge (\tilde{A} \geqslant \tilde{A}_k)] \\ &= \min V(\tilde{A} \geqslant \tilde{A}_i), i = 1, \ldots, k. \end{aligned} \quad (2.13)$$

For the triangular fuzzy numbers from Example 2.1, the following values of the respective degrees of possibility are obtained.

$$\begin{aligned} V(\tilde{A}_1 &\geqslant \tilde{A}_2, \tilde{A}_3, \tilde{A}_4) = 0.35, \\ V(\tilde{A}_2 &\geqslant \tilde{A}_1, \tilde{A}_3, \tilde{A}_4) = 0.96, \\ V(\tilde{A}_3 &\geqslant \tilde{A}_1, \tilde{A}_2, \tilde{A}_4) = 0.06, \\ V(\tilde{A}_4 &\geqslant \tilde{A}_1, \tilde{A}_2, \tilde{A}_3) = 1. \end{aligned}$$

This gives exactly the same order as the one obtained using the probabilistic approach and Yager index.

2.6 Weighted Averaging Approach Based on α–cuts

This section describes the ordering of *LR*-type fuzzy numbers associated with defuzzification of parametrically represented fuzzy numbers [23]. In the case of *LR*-type fuzzy numbers, the parametric representation (1.8) can be written in the following form:

$$\tilde{A} = \bigcap_{\alpha \in [0,1]} (\alpha, [L_{\tilde{A}}^{-1}(\alpha), R_{\tilde{A}}^{-1}(\alpha)]), \quad (2.14)$$

where $L_{\tilde{A}}^{-1}, R_{\tilde{A}}^{-1} : [0, 1] \rightarrow \Re$ are inverse functions of the respective shape functions of an *LR*-type fuzzy number $\tilde{A}$.

Definition 2.3 ([24]) Let $\tilde{A} \in \mathcal{F}(\Re)_{LR}$. The weighted averaging based on α–cuts representation of a fuzzy number $\tilde{A}$ is defined by:

$$I(\tilde{A}) = \int_0^1 (c_L L_{\tilde{A}}^{-1}(\alpha) + c_R R_{\tilde{A}}^{-1}(\alpha)) p(\alpha) d\alpha, \quad (2.15)$$

where c_L, c_R are, respectively, the optimism and pessimism parameters, $p(\alpha)$ is a distribution function of the importance of the α–cuts.

The $c_L; c_R$ parameters and the function $p(\alpha)$ satisfy the conditions:

$$c_L > 0, c_R > 0, c_L + c_R = 1,$$

$$p : [0, 1] \to \Re_+, \quad \int_0^1 p(\alpha)d\alpha = 1.$$

The function $p(\alpha)$ is also called the weighted averaging parameter. Following [23], it is assumed that

$$p(\alpha) = (k+1)\alpha^k,$$

where $k > 0$ is a parameter.

Theorem 2.1 ([24]) *Let $\tilde{A} = (a, b, \alpha, \beta)_{LR}$ and assume that the distribution of the function of the importance of the degrees have the form of relation (2.15). Then, the following formula is valid for weighted averaging:*

$$I(\tilde{A}) = c_L\left(\beta - \frac{k+1}{k+2}(\beta - \alpha)\right) + c_R\left(a - \frac{k+1}{k+2}(b-a)\right). \tag{2.16}$$

The value $I(\tilde{A})$ is a crisp value used to rank fuzzy numbers. The greater this value is, the greater is the fuzzy number. Moreover, $I(\tilde{A}) = I(\tilde{A}) = I(\tilde{B})$ if and only if $\tilde{A} = \tilde{B}$.

Example 2.2 In this example it is assumed that $p(\alpha) = 2$ ($k = 1$), and the "optimism/pessimism" coefficients are 0.5. Now, let the following three sets of trapezoidal fuzzy numbers and a set of triangular fuzzy numbers be given (see Fig. 2.4):

Set 1: $\tilde{A}_1 = (0.5, 0.5, 0.1, 0.5)$, $\tilde{A}_2 = (0.7, 0.7, 0.3, 0.3)$, $\tilde{A}_3 = (0.9, 0.9, 0.5, 0.1)$;
Set 2: $\tilde{A}_1 = (0.4, 0.7, 0.4, 0.1)$, $\tilde{A}_2 = (0.5, 0.5, 0.3, 0.4)$, $\tilde{A}_3 = (0.6, 0.6, 0.5, 0.2)$;
Set 3: $\tilde{A}_1 = (0.5, 0.5, 0.2, 0.2)$, $\tilde{A}_2 = (0.5, 0.8, 0.2, 0.1)$, $\tilde{A}_3 = (0.5, 0.5, 0.2, 0.4)$;
Set 4: $\tilde{A}_1 = (0.12, 0.19, 0.29)$, $\tilde{A}_2 = (0.22, 0.32, 0.48)$, $\tilde{A}_3 = (0.11, 0.15, 0.23)$, $\tilde{A}_4 = (0.21, 0.33, 0.49)$.

The ranking index values obtained for the set 1 are $I(\tilde{A}_1) = 0.37$, $I(\tilde{A}_2) = 0.50$, $I(\tilde{A}_3) = 0.63$. This gives the following order $\tilde{A}_1 < \tilde{A}_2 < \tilde{A}_3$. For the set 2, the ranking index values are $I(\tilde{A}_1) = 0.45$, $I(\tilde{A}_2) = 0.42$, $I(\tilde{A}_3) = 0.50$, which gives the order $\tilde{A}_2 < \tilde{A}_1 < \tilde{A}_3$. For the set 3, the ranking index values are $I(\tilde{A}_1) = 0.35$, $I(\tilde{A}_2) = 0.43$, $I(\tilde{A}_3) = 0.38$, which gives the order $\tilde{A}_1 < \tilde{A}_3 < \tilde{A}_2$. Finally, for the set 4 which is the same as in Example 2.1, the ranking index values are $I(\tilde{A}_1) = 0.14$, $I(\tilde{A}_2) = 0.22$, $I(\tilde{A}_3) = 0.10$, $I(\tilde{A}_4) = 0.23$. This gives exactly the same order as the one obtained using the previous approaches.

The α–cuts based approach is the most time consuming as it requires the pairwise comparison of all fuzzy numbers to be ordered. Also, the procedure of computing the possibility degree is more complicated that the computation of ranking indices in the two latter approaches.

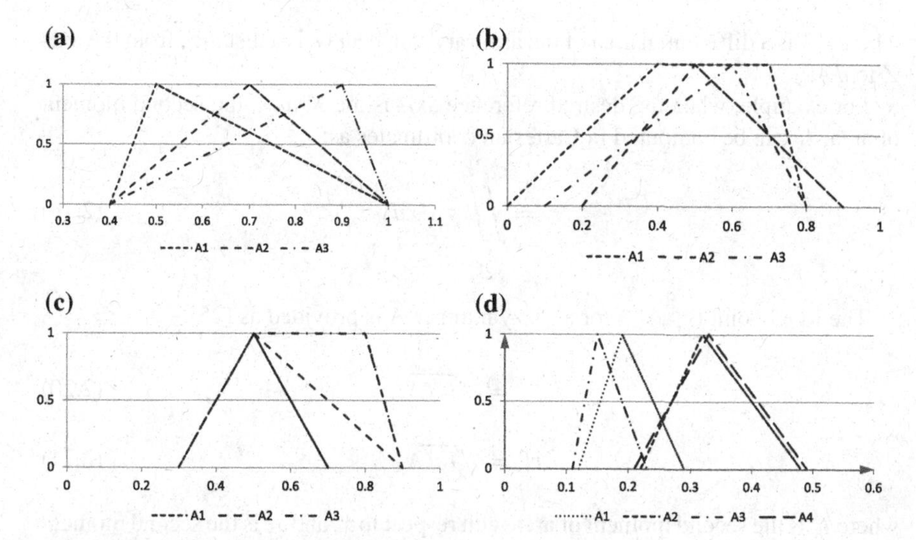

Fig. 2.4 Set 1 (**a**); Set 2 (**b**), Set 3 (**c**); Set 4 (**d**)

2.7 Two-Dimensional Radius of Gyration Approach

The two-dimensional radius of gyration (ROG) or *gyradius* is a concept in mechanics [25]:

$$r_g = \sqrt{I/A} \tag{2.17}$$

where I is the *second moment of area* (see Fig. 2.5) and A is the total cross-sectional area. The second moment of area of an arbitrary shape with respect to an arbitrary axis Z is defined and is computed by:

$$I_Z = \int_A r^2 dA, \tag{2.18}$$

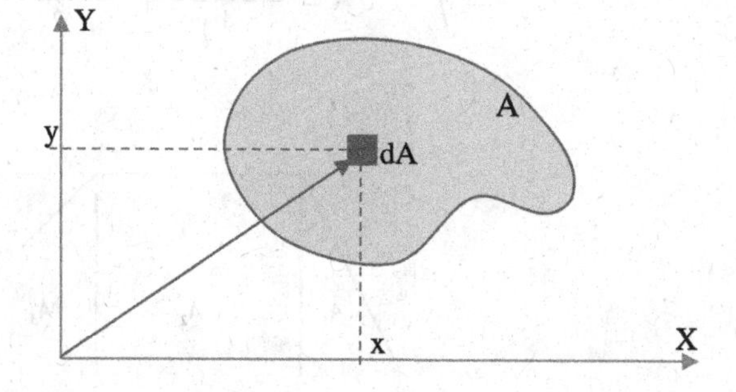

Fig. 2.5 A scheme of how the second moment of area is calculated for an arbitrary shape with respect to the Z axis; r is the radial distance to the element dA, with projections x and y on the axes

where dA is a differential area of the arbitrary shape and r is a distance from the axis Z to dA.

For example, when the desired reference axis is the X-axis, the second moment of area, I_x can be computed in Cartesian coordinates as:

$$I_x = \iint_A y^2 dxdy \tag{2.19}$$

The ROG point $(r_x^{\tilde{A}}, r_y^{\tilde{A}})$ for a fuzzy number $\tilde{A}$ is provided as [25]:

$$r_x^{\tilde{A}} = \sqrt{I_x/A}, \tag{2.20}$$

$$r_y^{\tilde{A}} = \sqrt{I_y/A}, \tag{2.21}$$

where I_x is the second moment of area with respect to x, and I_y is the second moment of area with respect to y. It is assumed that the mass density at each point of the area equals 1.

In the case of a trapezoidal fuzzy number, the second moment of area can be calculated in the following way. First, the trapezoidal area of a fuzzy number is divided onto three areas A_1, A_2, A_3 (see Fig. 2.6). It is known that for an area made up of a number of simple shapes, the second moment of area is the sum of the second moments of each of the individual areas about the desired axis [25]:

$$\begin{aligned} I_x &= I_x^{A_1} + I_x^{A_2} + I_x^{A_3} \\ I_y &= I_y^{A_1} + I_y^{A_2} + I_y^{A_3} \end{aligned} \tag{2.22}$$

The respective moments of inertia of the areas are given by [25]:

$$I_x^{A_1} = \int_{A_1} y^2 dA = \int_0^1 y^2(b-q)(1-y)dy = \frac{b-a}{12} \tag{2.23}$$

$$I_y^{A_1} = \int_{A_1} x^2 dA = \frac{(b-a)^3}{4} + \frac{(b-a)a^2}{2} + \frac{2(b-a)^2 a}{3} \tag{2.24}$$

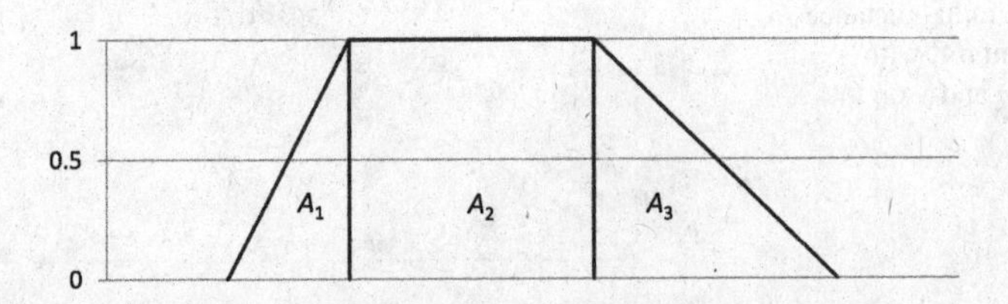

Fig. 2.6 A division of a trapezoid into three parts A_1, A_2 and A_3

$$I_x^{A_2} = \frac{(c-b)}{3} \tag{2.25}$$

$$I_y^{A_2} = \frac{(c-b)^3}{3} + (c-b)b^2 + (c-b)^2 b \tag{2.26}$$

$$I_x^{A_3} = \frac{(d-c)}{12} \tag{2.27}$$

$$I_y^{A_3} = \frac{(d-c)^3}{12} + \frac{(d-c)c^2}{2} + \frac{(d-c)^2 c}{3} \tag{2.28}$$

Then, the ROG point of a trapezoidal fuzzy number is calculated as [25]:

$$\begin{aligned} r_x^{\tilde{A}} &= \sqrt{\frac{I_x^{A_1}+I_x^{A_2}+I_x^{A_3}}{((c-b)+(d-a))/2}}, \\ r_y^{\tilde{A}} &= \sqrt{\frac{I_y^{A_1}+I_y^{A_2}+I_y^{A_3}}{((c-b)+(d-a))/2}}. \end{aligned} \tag{2.29}$$

For a crisp number a, the ROG point is defined by [25]:

$$r_x^a = \sqrt{3}/3, \quad r_y^a = a. \tag{2.30}$$

The ROG point $(r_x^{\tilde{A}}, r_y^{\tilde{A}})$ is used to define the index [25]

$$S(\tilde{A}) = r_x^{\tilde{A}} \cdot r_y^{\tilde{A}} \tag{2.31}$$

which is used to compare fuzzy numbers. The larger is the index, the greater is a fuzzy number. Thus, given two fuzzy numbers $\tilde{A}$ and $\tilde{B}$, the following holds:

$$\begin{aligned} &\text{If } S(\tilde{A}) > S(\tilde{B}) \text{ Then } \tilde{A} > \tilde{B}, \\ &\text{If } S(\tilde{A}) < S(\tilde{B}) \text{ Then } \tilde{A} < \tilde{B}, \\ &\text{If } S(\tilde{A}) = S(\tilde{B}) \text{ Then } \tilde{A} = \tilde{B}. \end{aligned} \tag{2.32}$$

Example 2.3 The values of the index S obtained for the fuzzy numbers from Example 2.1 are presented in Table 2.6.

This gives exactly the same ordering as those obtained using previously described approaches.

Table 2.6 Yager index for the fuzzy numbers from Example 2.1

Fuzzy number	S
$\tilde{A}_1 = (0.12, 0.19, 0.29)$	0.8167
$\tilde{A}_2 = (0.22, 0.32, 0.48)$	1.2484
$\tilde{A}_3 = (0.11, 0.15, 0.23)$	0.4410
$\tilde{A}_4 = (0.21, 0.33, 0.49)$	0.4217

2.8 Fuzzy Maximising-Minimising Points Approach

The ordering of fuzzy numbers using fuzzy maximising-minimising points is based on the centre of gravity point, defined in the previous section, left and right spreads and the distance between fuzzy numbers.

Definition 2.4 The distance between two arbitrary fuzzy numbers $\tilde{A}$ and $\tilde{B}$ is defined by:

$$d(\tilde{A}, \tilde{B}) = \left[\int_0^1 (f_{\tilde{A}}(y) - f_{\tilde{B}}(y))^2 dy + \int_0^1 (g_{\tilde{A}}(y) - g_{\tilde{B}}(y))^2 dy \right] \tag{2.33}$$

The fuzzy minimising-maximising points are obtained using the method from [26]. Let the fuzzy numbers $\tilde{A}_i$, $i = 1, 2, \ldots, n$ be given and let $\tilde{M}$ denote the fuzzy maximising point and $\tilde{m}$ the fuzzy minimising point. The *COGP* of the minimising and maximising points are computed as follows:

$$COGP(\tilde{M}) = \left(\max_{i=1,2,\ldots,n} \{\overline{x}_0(\tilde{A}_i)\}, \max_{i=1,2,\ldots n} \{\overline{y}_0(\tilde{A}_i)\} \right)$$

$$COGP(\tilde{m}) = \left(\min_{i=1,2,\ldots,n} \{\overline{x}_0(\tilde{A}_i)\}, \min_{i=1,2,\ldots n} \{\overline{y}_0(\tilde{A}_i)\} \right)$$

The left and right spreads of $\tilde{M}$ and $\tilde{m}$ are computed analogously:

$$L_{\tilde{M}} = \max_{i=1,2,\ldots,n} \{L_{\tilde{A}_i}\}, \quad R_{\tilde{M}} = \max_{i=1,2,\ldots,n} \{R_{\tilde{A}_i}\},$$

$$L_{\tilde{m}} = \min_{i=1,2,\ldots n} \{L_{\tilde{A}_i}\}, \quad R_{\tilde{m}} = \min_{i=1,2,\ldots n} \{R_{\tilde{A}_i}\}.$$

Now, given $COGP(\tilde{M})$, $L_{\tilde{M}}$, $R_{\tilde{M}}$ and $COGP(\tilde{m})$ $L_{\tilde{m}}$, $R_{\tilde{m}}$, the goal is to uniquely determine, respectively, $\tilde{M}$ and $\tilde{m}$.

In general case, an unknown fuzzy number $\tilde{A}$ can be uniquely determined based on its centroid point $(\overline{x}_0(\tilde{A}), \overline{y}_0(\tilde{A}))$ and left L and right R spreads by solving the following system of nonlinear equations [26]:

$$\begin{cases} b - a = L \\ d - c = R \\ \frac{1}{3}\left[a + b + c + d - \frac{cd-ab}{(c+d)-(a+b)}\right] = \overline{x}_0(\tilde{A}) \\ \frac{1}{3}\left[1 + \frac{c-b}{(c+d)-(a+b)}\right] = \overline{y}_0(\tilde{A}) \end{cases}$$

The fuzzy ranking using fuzzy minimising-maximising points uses the following relative closeness coefficient:

$$D(\tilde{A}) = \gamma(\tilde{A}) \cdot \frac{D^L_{\tilde{A}}}{1 + D^R_{\tilde{A}}}, \tag{2.34}$$

where

$$D^L_{\tilde{A}} = d(\tilde{A}, \tilde{m}),$$

$$D^R_{\tilde{A}} = d(\tilde{A}, \tilde{M}),$$

and

$$\gamma(\tilde{A}) = \begin{cases} 1, & \text{if } \int_0^1 \{f_{\tilde{A}}(y) + g_{\tilde{A}}(y)\}dy \geqslant 0, \\ -1, & \text{if } \int_0^1 \{f_{\tilde{A}}(y) + g_{\tilde{A}}(y)\}dy < 0. \end{cases}$$

The ranking rules for fuzzy numbers $\tilde{A}_i$, $i = 1, 2, \ldots, n$, are the following [26]:

$$\begin{array}{l} A_i < A_j \text{ iff } D(A_i) < D(A_j) \\ A_i > A_j \text{ iff } D(A_i) > D(A_j) \\ A_i \approx A_j \text{ iff } D(A_i) = D(A_j) \end{array}$$

Example 2.4 Consider the fuzzy numbers from Example 2.1. The corresponding fuzzy maximising and minimising points are depicted in Fig. 2.7.

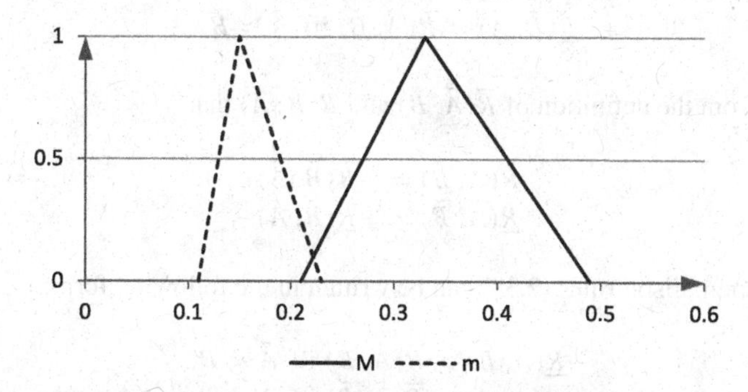

Fig. 2.7 Fuzzy maximising $\tilde{M}$ and minimising $\tilde{m}$ points

The following values of the relative closeness coefficient were obtained: $D(A_1) = 0.0067$, $D(A_2) = 0.0395$, $D(A_3) = 0.0015$, $D(A_4) = 0.0468$, which gives exactly the same ordering as those obtained using previously described approaches.

The method is quite complicated compared to other presented methods. It is also time consuming and requires to use additional tools, such as numerical integration and solving systems of nonlinear equations. It also requires the pairwise comparison of fuzzy numbers to be ordered.

2.9 Area Based Approach

For a fuzzy number $\tilde{A} = (f_{\tilde{A}}, g_{\tilde{A}})$, the following values are defined [27]:

$$\begin{aligned} \tilde{A}_l &= m + \tfrac{1}{2} H_l, \\ \tilde{A}_u &= m + \tfrac{1}{2} H_u \end{aligned} \tag{2.35}$$

where $m = \frac{1}{2}(f_{\tilde{A}}^{-1}(1) + g_{\tilde{A}}^{-1}(1))$, and H_l, H_u are defined as follows:

$$\begin{aligned} H_l &= \frac{\int_0^1 f_{\tilde{A}}(y)dy}{\int_0^1 f_{\tilde{A}}(y)dy + \int_0^1 g_{\tilde{A}}(y)dy}, \\ H_u &= \frac{\int_0^1 g_{\tilde{A}}(y)dy}{\int_0^1 f_{\tilde{A}}(y)dy + \int_0^1 g_{\tilde{A}}(y)dy} \end{aligned} \tag{2.36}$$

Now, for given two fuzzy numbers $\tilde{A}$ and $\tilde{B}$, the following values are defined [27]:

$$\overline{R}(\tilde{A}, \tilde{B}) = \tilde{A}_u - \tilde{B}_u, \quad \underline{R}(\tilde{A}, \tilde{B}) = \tilde{A}_l - \tilde{B}_l \tag{2.37}$$

They are used to determine the comparison rules:

$$\begin{aligned} \underline{R}(\tilde{B}, \tilde{A}) &> \overline{R}(\tilde{A}, \tilde{B}) \text{ iff } \tilde{A} < \tilde{B}, \\ \underline{R}(\tilde{B}, \tilde{A}) &= \overline{R}(\tilde{A}, \tilde{B}) \text{ iff } \tilde{A} \approx \tilde{B}. \end{aligned} \tag{2.38}$$

It follows from the definition of $\overline{R}(\tilde{A}, \tilde{B})$ and $\underline{R}(\tilde{B}, \tilde{A})$ that

$$\begin{aligned} \overline{R}(\tilde{A}, \tilde{B}) &= -\overline{R}(\tilde{B}, \tilde{A}) \\ \underline{R}(\tilde{A}, \tilde{B}) &= -\underline{R}(\tilde{B}, \tilde{A}) \end{aligned}$$

Thus, the comparison rules (2.38) can be written in the following form:

$$\begin{aligned} -\underline{R}(\tilde{A}, \tilde{B}) &> \overline{R}(\tilde{A}, \tilde{B}) \text{ iff } \tilde{A} < \tilde{B}, \\ -\underline{R}(\tilde{A}, \tilde{B}) &= \overline{R}(\tilde{A}, \tilde{B}) \text{ iff } \tilde{A} = \tilde{B}. \end{aligned} \tag{2.39}$$

Table 2.7 The values of $\underline{R}$, $\overline{R}$ and the pairwise comparison results

$\underline{R}$	$\overline{R}$	Comparison
$\underline{R}(\tilde{A}_1, \tilde{A}_2) = -0.07558$	$\underline{R}(\tilde{A}_1, \tilde{A}_2) = -0.05442$	$\tilde{A}_1 < \tilde{A}_2$
$\underline{R}(\tilde{A}_1, \tilde{A}_3) = 0.006155$	$\underline{R}(\tilde{A}_1, \tilde{A}_3) = 0.033845$	$\tilde{A}_1 > \tilde{A}_3$
$\underline{R}(\tilde{A}_1, \tilde{A}_4) = -0.074654$	$\underline{R}(\tilde{A}_1, \tilde{A}_4) = -0.065346$	$\tilde{A}_1 < \tilde{A}_4$
$\underline{R}(\tilde{A}_2, \tilde{A}_3) = 0.081735$	$\underline{R}(\tilde{A}_2, \tilde{A}_3) = 0.088265$	$\tilde{A}_2 > \tilde{A}_3$
$\underline{R}(\tilde{A}_2, \tilde{A}_4) = 0.000926$	$\underline{R}(\tilde{A}_2, \tilde{A}_4) = -0.010926$	$\tilde{A}_2 > \tilde{A}_4$
$\underline{R}(\tilde{A}_3, \tilde{A}_4) = -0.080809$	$\underline{R}(\tilde{A}_3, \tilde{A}_4) = -0.099191$	$\tilde{A}_3 > \tilde{A}_4$

Example 2.5 Consider the fuzzy numbers from Example 2.1. The values of $\overline{R}$, $\underline{R}$ and the pairwise comparison results are presented in Table 2.7.

This gives exactly the same order as the one obtained using the methods described so far.

2.10 Left and Right Dominance Approach

The ordering of fuzzy numbers based on the left and right dominance was proposed in [28]. This approach uses left and right bounds of selected α–cuts of fuzzy numbers to be compared.

Definition 2.5 The left $D^L_{i,j}$ and right $D^R_{i,j}$ dominance of a fuzzy number $\tilde{A}_i$ over a fuzzy number $\tilde{A}_j$ is defined as the average difference of the left and right bounds of $\tilde{A}_i$ and $\tilde{A}_j$ at some α-levels:

$$D^L_{ij} = \frac{1}{n+1} \sum\nolimits_{k=0}^{n} (l_{ik} - l_{jk}) \tag{2.40}$$

$$D^R_{ij} = \frac{1}{n+1} \sum\nolimits_{k=0}^{n} (r_{ik} - r_{jk}) \tag{2.41}$$

where n is the numbers of α–cuts, l_{ik}, r_{ik} are, respectively, left and right spreads of a fuzzy number $\tilde{A}_i$ at the α_k-level.

It is assumed that α-levels are spread uniformly, i.e., k/n, $k = 1, 2, \ldots, n$. The values $D^L_{i,j}$ and $D^R_{i,j}$ approximate the area difference of $\tilde{A}_i$ over $\tilde{A}_j$ according to the membership axis to the, respectively, left and right membership function as $n \to \infty$ [28]. The total dominance of $\tilde{A}_i$ over $\tilde{A}_j$ with the index of optimism $\beta \in [0, 1]$ is defined as follows.

Definition 2.6 The total dominance $D^T_{i,j}$ of a fuzzy number $\tilde{A}_i$ over a fuzzy number $\tilde{A}_j$ with optimism index $\beta \in [0, 1]$ is defined as a convex combinations of left $D^L_{i,j}$

Table 2.8 The values of total dominance

Total dominance index	Value
$D_{12}^T(0.5)$	−0.1375
$D_{13}^T(0.5)$	0.0375
$D_{14}^T(0.5)$	−0.1425
$D_{23}^T(0.5)$	0.1750
$D_{24}^T(0.5)$	−0.005
$D_{34}^T(0.5)$	−0.1800

and right $D_{i,j}^R$ dominance:

$$D_{ij}^T(\beta) = \beta D_{ij}^L + (1-\beta)D_{ij}^R \tag{2.42}$$

The index of optimism is used to reflect a decision maker's degree of optimism [28].

The total dominance index is used to define the rules of comparison of two fuzzy numbers. They are the following:

$$\begin{aligned}&\text{If } D_{ij}^T < 0 \text{ Then } \tilde{A}_i < \tilde{A}_j,\\ &\text{If } D_{ij}^T > 0 \text{ Then } \tilde{A}_i > \tilde{A}_j,\\ &\text{If } D_{ij}^T = 0 \text{ Then } \tilde{A}_i = \tilde{A}_j.\end{aligned}$$

Example 2.6 Consider the fuzzy numbers from Example 2.1. The obtained values of the total dominance with $\beta = 0.5$ and $n = 5$ are summarised in Table 2.8.

This gives the ordering $\tilde{A}_3 < \tilde{A}_1 < \tilde{A}_2 < \tilde{A}_4$, which exactly the same as the one obtained using the previous methods.

2.11 An α-weighted Valuations Approach

Approaches to the ranking of fuzzy numbers based upon the idea of associating with a fuzzy number a scalar value, i.e., its valuation, was developed by Yager [22]. Later, Yager and Filev [29] improved this valuation method by the transformation of a fuzzy subset into an associated probability distribution. They introduced a family of parametric valuation functions. The problem of ranking fuzzy numbers using valuation methods was also considered by Detyniecki and Yager [3].

A generalised formula for a class of valuation functions has the form:

$$Val(\tilde{A}) = \frac{\int_0^1 Ave(\tilde{A}^\alpha) f(\alpha) d\alpha}{\int_0^1 f(\alpha) d\alpha} \tag{2.43}$$

where f is a mapping $f : [0, 1] \rightarrow [0, 1]$. In [29], Yager and Filev proposed two complementary families of parametric valuation functions. The one is an increasing family:

$$f : [0, 1] \ni \alpha \rightarrow \alpha^q \in [0, 1], \quad q > 0,$$

and the other one is decreasing

$$f : [0, 1] \ni \alpha \rightarrow (1 - \alpha)^q \in [0, 1], \quad q > 0.$$

Some interesting properties of this two families of functions can be found in [29]. One of them is that increasing family emphasises the higher α-levels, whereas the decreasing family emphasises lower α-levels, which causes that these two families can produce two opposite orderings.

In order to calculate the valuation for a given fuzzy number $\tilde{A}$, the value of

$$Ave(\tilde{A}^\alpha) = \frac{\inf(\tilde{A}_\alpha) + \sup(\tilde{A}_\alpha)}{2}$$

must be first computed. In the case of trapezoidal fuzzy numbers

$$Ave(\tilde{A}^\alpha) = \frac{a + (b - a)\alpha + d - (d - c)\alpha}{2} = \frac{b + c}{2}\alpha + \frac{a + d}{2}(1 - \alpha).$$

Then, the valuation formula (2.43) takes the form:

$$Val(\tilde{A}) = \frac{\frac{1}{2}\int_0^1 ((b + c)\alpha + (a + d)(1 - \alpha))f(\alpha)d\alpha}{\int_0^1 f(\alpha)d\alpha}. \quad (2.44)$$

which can be simplified to:

$$Val(\tilde{A}) = \frac{b + c}{2}w + \frac{a + d}{2}(1 - w),$$

where

$$w = \frac{\int_0^1 \alpha f(\alpha)d\alpha}{\int_0^1 f(\alpha)d\alpha}.$$

For the increase case

$$w = \frac{q + 1}{q + 2},$$

and for the decrease case

$$w = \frac{1}{q + 2}.$$

Table 2.9 The results of comparison of $\tilde{A}_1$ and $\tilde{A}_2$ using increasing valuation function with different values of the parameter q

q	$Val(\tilde{A}_1)$	$Val(\tilde{A}_2)$	Comparison
0	6	7	$<$
1	6.33	6.67	$<$
2	6.5	6.5	$=$
3	6.6	6.4	$>$
∞	7	6	$>$

Example 2.7 Consider the following triangular fuzzy numbers: $\tilde{A}_1 = (1, 7, 9)$, $\tilde{A}_2 = (4, 6, 12)$, $\tilde{A}_3 = (5, 8, 9)$, $\tilde{A}_4 = (2, 9, 10)$. The results of comparison of $\tilde{A}_1$ and $\tilde{A}_2$, obtained using an increasing functions with different values of q, are presented in Table 2.9. The results of ordering obtained using increasing and decreasing functions with $q = 2$ are given in Table 2.11. Finally, Table 2.10 presents the ordering of fuzzy numbers from Example 2.1 obtained using decreasing and increasing valuation functions with different value of the parameter q.

The results show that for $q = 2$, the ordering of fuzzy numbers is exactly the same as the one obtained using other considered methods.

Table 2.10 The results of comparison of exemplary fuzzy numbers using increasing and decreasing valuation functions with different values of the parameter q

	$Val(\cdot)$ (increasing)	$Val(\cdot)$ (decreasing)	Order (increasing)	Order (decreasing)
$q = 0$				
$\tilde{A}_1$	6.0	6.0	$\tilde{A}_4 = \tilde{A}_3 > \tilde{A}_2 > \tilde{A}_1$	$\tilde{A}_4 = \tilde{A}_3 > \tilde{A}_2 > \tilde{A}_1$
$\tilde{A}_2$	7.0	7.0		
$\tilde{A}_3$	7.5	7.5		
$\tilde{A}_4$	7.5	7.5		
$q = 1$				
$\tilde{A}_1$	6.33	5.67	$\tilde{A}_4 > \tilde{A}_3 > \tilde{A}_2 > \tilde{A}_1$	$\tilde{A}_3 = \tilde{A}_2 > \tilde{A}_4 > \tilde{A}_1$
$\tilde{A}_2$	6.67	7.33		
$\tilde{A}_3$	7.67	7.33		
$\tilde{A}_4$	8.00	7.00		
$q = 2$				
$\tilde{A}_1$	6.5	5.5	$\tilde{A}_4 > \tilde{A}_3 > \tilde{A}_2 = \tilde{A}_1$	$\tilde{A}_2 > \tilde{A}_3 > \tilde{A}_4 > \tilde{A}_1$
$\tilde{A}_2$	6.5	7.5		
$\tilde{A}_3$	7.75	7.25		
$\tilde{A}_4$	8.25	6.75		
$q = \infty$				
$\tilde{A}_1$	7.0	5.0	$\tilde{A}_4 > \tilde{A}_3 > \tilde{A}_1 > \tilde{A}_2$	$\tilde{A}_2 > \tilde{A}_3 > \tilde{A}_4 > \tilde{A}_1$
$\tilde{A}_2$	6.0	8.0		
$\tilde{A}_3$	8.0	7.0		
$\tilde{A}_4$	9.0	6.0		

Table 2.11 The results of comparison of exemplary fuzzy numbers using increasing and decreasing valuation functions with different values of the parameter q

	$Val(\cdot)$ (increasing)	$Val(\cdot)$ (decreasing)	Order (increasing)	Order (decreasing)
$q = 0$				
$\tilde{A}_1$	0.2	0.2	$\tilde{A}_4 = \tilde{A}_2 > \tilde{A}_1 > \tilde{A}_3$	$\tilde{A}_4 = \tilde{A}_2 > \tilde{A}_1 > \tilde{A}_3$
$\tilde{A}_2$	0.34	0.34		
$\tilde{A}_3$	0.16	0.16		
$\tilde{A}_4$	0.34	0.34		
$q = 1$				
$\tilde{A}_1$	0.2	0.2	$\tilde{A}_4 > \tilde{A}_2 > \tilde{A}_1 > \tilde{A}_3$	$\tilde{A}_4 = \tilde{A}_2 > \tilde{A}_1 > \tilde{A}_3$
$\tilde{A}_2$	0.33	0.34		
$\tilde{A}_3$	0.16	0.16		
$\tilde{A}_4$	0.34	0.34		
$q = 2$				
$\tilde{A}_1$	0.19	0.2	$\tilde{A}_4 > \tilde{A}_2 > \tilde{A}_1 > \tilde{A}_3$	$\tilde{A}_4 > \tilde{A}_2 > \tilde{A}_1 > \tilde{A}_3$
$\tilde{A}_2$	0.33	0.34		
$\tilde{A}_3$	0.16	0.17		
$\tilde{A}_4$	0.34	0.35		
$q = \infty$				
$\tilde{A}_1$	0.19	0.2	$\tilde{A}_4 = \tilde{A}_2 > \tilde{A}_1 > \tilde{A}_3$	$\tilde{A}_4 = \tilde{A}_2 > \tilde{A}_1 > \tilde{A}_3$
$\tilde{A}_2$	0.33	0.35		
$\tilde{A}_3$	0.15	0.17		
$\tilde{A}_4$	0.33	0.35		

References

1. Baas, S.M., and H. Kwakernaak. 1977. Rating and ranking of multiple-aspect alternative using fuzzy sets. *Automatica* 13(1): 47–58.
2. Chanas, S., and D. Kuchta. 1996. Multiobjective programming in optimization of interval objective functions a generalized approach. *European Journal of Operational Research* 94(3): 594–598.
3. Detyniecki, M., and R.R. Yager. 2000. Ranking fuzzy numbers using α-weighted valuations. *International Journal of Uncertainty, Fuzziness and Knowledge-Based Systems* 8(3): 573–591.
4. Dubois, D., and H. Prade. 1983. Ranking of fuzzy numbers in the setting of possibility theory. *Information Sciences* 30(3): 183–224.
5. Sengupta, A., and T.P. Kumar. 2009. Fuzzy preference ordering of intervals. *Fuzzy preference ordering of interval numbers in decision problems*, Studies in fuzziness and soft computing, vol. 238, pp. 59–89. Berlin: Springer.
6. Yager, R.R., M. Detyniecki, and B. Bouchon-Meunier. 2001. A context-dependent method for ordering fuzzy numbers using probabilities. *Information Sciences* 30(3): 237–255.
7. Jain, R. 1976. Decision-making in the presence of fuzzy variables. *IEEE Transactions on Systems, Man and Cybernetics* 6(10): 698–703.
8. Bortolan, G., and R. Degani. 1985. A review of some methods for ranking fuzzy subsets. *Fuzzy Sets and Systems* 15(1): 1–19.

9. Kim, K.T., and D.H. Hong. 2006. An easy computation of min and max operations for fuzzy numbers. *Fuzzy Sets and Systems* 21(1–2): 555–561.
10. Asady, B., and A. Zendehnam. 2007. Ranking fuzzy numbers by distance minimization. *Applied Mathematical Modelling* 31(11): 2589–2598.
11. Thorani, Y.L.P., P.P.B. Rao, and N.R. Shankar. 2012. Ordering generalized trapezoidal fuzzy numbers. *International Journal of Contemporary Mathematical Sciences* 7(12): 555–573.
12. Allahviranloo, T., M.A. Jahantigh, and S. Hajighasemi. 2013. A new distance measure and ranking method for generalized trapezoidal fuzzy numbers. *Mathematical Problems in Engineering* 2013: 1–6.
13. Sevastjanov, P.V., and P. Rog. 2003. A probabilistic approach to fuzzy and crisp interval ordering. *TASK Quarterly : Scientific bulletin of Academic Computer Centre in Gdansk* 7(1): 147–156.
14. Huynh, V.-N., Y. Nakamori, and J. Lawry. 2008. A probability-based approach to comparison of fuzzy numbers and applications to target-oriented decision making. *IEEE Transactions on Fuzzy Systems* 16(2): 371–387.
15. Moore, R.E. 1966. *Interval analysis.*, Prentice-Hall, Series in automatic computation New Jersey: Prentice-Hall.
16. Jiang, C., X. Han, G.R. Liu, and G.P. Liu. 2008. A nonlinear interval number programming method for uncertain optimization problems. *European Journal of Operational Research* 188(1): 1–13.
17. Sevastianov, P., P. Rog, and K. Karczewski. 2002. A probabilistic method for ordering group of intervals. *Informatyka Teoretyczna i Stosowana* R2(2): 45–53.
18. Wang, X., and E.E. Kerre. 2001. Reasonable properties for the ordering of fuzzy quantities (ii). *Fuzzy Sets and Systems* 118(3): 387–405.
19. Chang, D.-Y. 1996. Applications of the extent analysis method on fuzzy AHP. *European Journal of Operational Research* 95(3): 649–655.
20. Wang, Y.M., J.B. Yang, D.L. Xu, and K.S. Chin. 2006. On the centroids of fuzzy numbers. *Fuzzy Sets and Systems* 157(7): 919–926.
21. Abdullah, L., and N.J. Jamal. 2010. Centroid-point of ranking fuzzy numbers and its application to health related quality of life indicators. *International Journal on Computer Science and Engineering* 2(8): 2773–2777.
22. Yager, R. 1981. A procedure for ordering fuzzy subsets of the unit interval. *Information Sciences* 24(2): 143–161.
23. Saneifard, R. 2009. Ranking L-R fuzzy numbers with weighted averaging based on levels. *International Journal of Industrial Mathematics* 1(2): 163–173.
24. Nasibov, E.N., and A. Mert. 2007. On methods of defuzzification of parametrically represented fuzzy numbers. *Automatic Control and Computer Sciences* 41(5): 163–173.
25. Deng, Y., Z. Zhenfu, and L. Qi. 2006. Ranking fuzzy numbers with an area method using radius of gyration. *Computers and Mathematics with Applications* 51(6–7): 1127–1136.
26. Salahshour, S., S. Abbasbandy, and T. Allahviranloo. 2011. Ranking fuzzy numbers using fuzzy maximizing-minimizing points. In *2011.115, Proceedings of the 7th conference of the European Society for Fuzzy Logic and Technology (EUSFLAT-2011) and LFA-2011*, ed. Galichet, S., J. Montero, and G. Mauris, Advances in intelligent systems research, vol. 1, pp. 763–769. European Society for Fuzzy Logic and Technology (EUSFLAT-2011) and LFA.
27. Nasseri, S.H., F. Taleshian, Z. Alizadeh, and J. Vahidi. 2012. A new method for ordering LR fuzzy numbern. *The Journal of Mathematics and Computer Science* 4(3): 283–294.
28. Chen, L.-H., and H.-W. Lu. 2001. An approximate approach for ranking fuzzy numbers based on left and right dominance. *Computers and Mathematics with Applications* 41(12): 1589–1602.
29. Yager, R.R., and D. Filev. 1999. On ranking fuzzy numbers using valuations. *International Journal of Intelligent Systems* 14(12): 1249–1268.

Chapter 3
Fuzzy Random Variable and the Dempster-Shafer Theory of Evidence

Abstract This chapter presents the concept of uncertainty propagation in real world desicion problems, where some input parameters are stochastic while information about others is partial and is represented by fuzzy random variable. It also introduces fuzzy random variable and the Dempster-Shafer theory which provide mathematical background for such propagation.

3.1 Fuzzy Random Variable

In many decision problems, it is necessary to jointly consider randomness and imprecision, because data coming from various sources can be subject to both types of uncertainty. One of the possible approaches that enables to deal with this issue is based on the concept of a *fuzzy random variable*, which extends the classical definition of a random variable. It was introduced by Féron [1] and later modified by, e.g., Nahmias [2], Stein and Talati [3], Kwakernaak [4, 5], Puri and Ralescu [6], Diamond and Kloeden [7] and Kruse [8]. Krätschmer [9] surveyed all of these definitions and proposed a unified approach. He defined a fuzzy random variable as a function (mapping) that assigns a fuzzy subset to each of possible outputs of a random experiment. The similar approach to a fuzzy random variable was proposed by Liu and Liu [10–12]. Their approach is adopted in this book.

Definition 3.1 Let $(\Omega, \Sigma, \mathrm{P})$ be a probability space. A fuzzy random vector is a mapping $\xi = (\xi_1, \xi_2, \ldots, \xi_n) : \Omega \to \mathcal{F}^n(\Re)$ such that for any closed $C \subseteq \Re^n$

$$\xi^*(C)(\omega) = Pos\{\xi(\omega) \in C\} = \sup_{t \in C} \mu_{\xi(\omega)}(t) \tag{3.1}$$

is a measurable function of ω, where $\mu_{\xi(\omega)}$ is defined as

$$\mu_{\xi(\omega)}(t) = \min_{1 \leqslant i \leqslant} \mu_{\xi_i(\omega)}(t_i) \tag{3.2}$$

for any $t = (t_1, t_2, \ldots, t_n) \in \Re^n$.

© Springer International Publishing Switzerland 2015
I. Skalna et al., *Advances in Fuzzy Decision Making*,
Studies in Fuzziness and Soft Computing 333,
DOI 10.1007/978-3-319-26494-3_3

Three kinds of common fuzzy random variables, triangular, trapezoidal, and normal are used the most often. They are defined as follows:

- a fuzzy random variable ξ is said to be triangular if for each ω, $\xi(\omega)$ is a triangular fuzzy variable given by $(a(\omega), b(\omega), c(\omega))$, where a, b and c are random variables defined on a probability space Ω,
- a fuzzy random variable ξ is said to be trapezoidal if for each ω, $\xi(\omega)$ is a trapezoidal fuzzy variable given by $(a(\omega), b(\omega), c(\omega), d(\omega))$, where a, b, c and d are random variables defined on a probability space Ω,
- a fuzzy random variable ξ is said to be normal if for each ω, $\xi(\omega)$ is a fuzzy variable with a membership function given by

$$\mu(r) = \exp\left(-\left(\frac{r - c(\omega)}{w(\omega)}\right)^2\right)$$

where $c(\omega)$ and $w(\omega)$ are random variables defined on a probability space Ω.

3.2 Dempster–Shafer (D–S) Theory of Evidence

The theory of *belief functions* (also called *evidence theory*) was introduced by Shafer [13]. It allows imprecision and variability to be treated separately within a single framework. Indeed, belief functions provide mathematical tools to process information, which is at the same time of random and imprecise nature. This kind of knowledge is typically found in real decision processes, where some parameters are described by probability distributions, whereas others are described by fuzzy numbers. Contrary to probability theory, which assigns probability weights to atoms (elements of the referential), the theory of evidence may assign such weights to any subsets, called focal sets, with the understanding that portions of these weights may move freely from one element of such subsets to another. Most often, a sample of random intervals is obtained. In such case, information is presented in the form of intervals $[\underline{a}_i, \overline{a}_i]$ $(i = 1, 2, \ldots, I)$. To each interval is attached a probability v_i. That is, a mass distribution v_i on intervals is obtained. The probability mass v_i can be freely re-allocated to points within interval $[\underline{a}_i, \overline{a}_i]$. However, there is not enough information to do it.

As in the possibility theory, the evidence theory provides two indicators, plausibility *Pl* and belief *Bel* to qualify the validity of a proposition stating that the value of variable X should lie within a set A (a certain interval for example). Plausibility *Pl* and belief *Bel* measures are defined from the *mass distribution*.

Definition 3.2 Let $P(\Omega)$ be the power set of Ω. Then, the mass distribution is a mapping

$$v : P(\Omega) \to [0, 1], \tag{3.3}$$

such that $\sum_{E \in P(\Omega)} v(E) = 1$.

Definition 3.3 The belief function is defined by:

$$Bel(A) = \sum_{E,\, E \subseteq A} v(E). \tag{3.4}$$

Definition 3.4 The plausibility function is defined by:

$$Pl(A) = \sum_{E,\, E \cap A \neq \varphi} v(E) = 1 - Bel(\overline{A}). \tag{3.5}$$

The element E is called a focal element of $P(\Omega)$ if $v(E) > 0$.

$Bel(A)$ gathers the imprecise evidence that asserts A. Following Dempster [14], this is the minimal amount of probability that can be assigned to A by sharing the probability weights defined by the mass function among single values in the focal sets. Whereas $Pl(A)$ gathers the imprecise evidence that does not contradict A. This is the maximal amount of probability that can be assigned to A in the same fashion.

Evidence theory encompasses possibility and probability theory because [13, 15]:

- When focal elements are nested, the belief measure *Bel* is a necessity measure, i.e., $Bel = Nec$, and a plausibility measure *Pl* is a possibility measure, i.e., $Pl = Pos$.
- When focal elements are some disjoint intervals, the plausibility *Pl* and belief *Bel* measures are both probability measures, i.e., $Bel = P = Pl$, for unions of intervals.

Thus, all probability distributions and all possibility distributions may be interpreted as mass functions. Hence, one may work in a common framework to treat information of imprecise and random nature.

Below are presented the ways of creating the probability mass on the basis of the probability distribution and the possibility distribution.

- Probability → Density function
  Let X be a real random variable with a probability density p_X. By discretizing X into m disjoint intervals $([a_i,\ a_{i+1}])_{i=1,2,\ldots,m}$ being used to define focal elements, the mass distribution $(v_i)_{i=1,2,\ldots,m}$ can be built as follows:

$$v([a_i,\ a_{i+1}]) = v_i = P(X \in [a_i,\ a_{i+1}])$$

- Possibility → Belief function
  Let Y be a possibilistic variable. The possibility distribution of Y is denoted here by π, and π^{α} denotes an α-cut of π. Focal elements for Y corresponding to α-cuts are denoted by $(\pi^{\alpha_j})_{j=1,\ldots,q}$ with $\alpha_0 = \alpha_1 = 1 > \alpha_2 > \cdots \alpha_q > \alpha_{q+1} = 0$. Finally, the mass distribution associated to $(\pi^{\alpha_j})_{j=1,\ldots,q}$ is denoted by $(v_j = \alpha_j - \alpha_{j+1})_{j=1,\ldots,q}$.

Any pair of functions [*Nec, Pos*] or [*Bel, Pl*] can be interpreted as upper and lower probabilities induced from specific probability families. As stated above, every possibility distribution π induces a pair of functions *Nec and Pos*. The probability family is defined by

$$\mathcal{P}(\pi) = \{P, \forall A,\ Nec(A) \leqslant P(A)\} = \{P, \forall A,\ P(A) \leqslant Nec(A)\}$$

In this case $\sup_{P \in P(\pi)} P(A) = Pos(A)$ and $\inf_{P \in P(\pi)} P(A) = Nec(A)$, and thus $\overline{P} = Pos$ and $\underline{P} = Nec$. Hence, upper $\overline{F}$ and lower $\underline{F}$ cumulative distribution functions can be defined such that $\forall x \in \Re\ \underline{F}(x) \leqslant F(x) \leqslant \overline{F}(x)$ where:

$$\overline{F}(x) = Pos(X \in [-\infty, x]), \tag{3.6}$$

$$\underline{F}(x) = Nec(X \in [-\infty, x]). \tag{3.7}$$

Similarly a mass distribution v may encode probability family $\mathcal{P}(v) = \{P, \forall A,$ $Bel(A) \leqslant P(A)\} = \{P, \forall A,\ P(A) \leqslant Pl(A)\}$. In this case $\overline{P} = Pl$ and $\underline{P} = Bel$. Hence, upper $\overline{F}$ and lower $\underline{F}$ can be defined as follows: $\forall x$

$$\overline{F}(x) = Pl(X \in [-\infty, x]), \tag{3.8}$$

$$\underline{F}(x) = Bel(X \in [-\infty, x]). \tag{3.9}$$

3.3 The Hybrid Data Propagation Method

This section describes a method for processing of hybrid data, i.e., data consisting of both random and fuzzy variables. The method aims to determine the value of $\hat{f}(\hat{X})$, where $\hat{X} = (X_1, X_2, \ldots, X_m)$ is a vector of variables burdened with uncertainty. It is assumed that there are k ($k < m$) random variables $(X_1, X_2, \ldots, X_k)$ and $m - k$ fuzzy variables $\tilde{X}_{k+1}, \tilde{X}_{k+2}, \ldots, \tilde{X}_m$. Additionally, it is assumed that there may be defined subsets X^K of correlated variables X_i; $X^K = \{X_i \mid i \in K\}$, $K \in K_s$. In such case, K is the subset of correlated variable indices, and K_s is the set of indices of the selected subsets of correlated variables.

The proposed method (see Algorithm 1) for determining the value $\hat{f}(\hat{X})$ combines stochastic simulation with nonlinear programming. The latter is used to take into account dependencies between fuzzy variables. The computational procedure performs as follows. Values $(x_1, x_2, \ldots, x_k)$ of random variables are drawn using a procedure which accounts for correlation of the stochastic variables. The obtained values and fuzzy variables $\tilde{X}_{k+1}, \tilde{X}_{k+2}, \ldots, \tilde{X}_m$ allow to determine $\hat{f}(x_1, x_2, \ldots, x_k, \tilde{X}_{k+1}, \tilde{X}_{k+2}, \ldots, \tilde{X}_m)$ as a fuzzy number. This can be achieved using the concept of α-cuts. Upper bound (*sup*) and lower bound (*inf*) of an α-cut of a fuzzy number $\hat{f}(x_1, x_2, \ldots, x_k, \tilde{X}_{k+1}, \tilde{X}_{k+2}, \ldots, \tilde{X}_m)$ may be determined by solving the following non-linear programming tasks.

When searching for supremum, find:

$$f\left(\mathrm{x}_1, \mathrm{x}_2, \ldots, \mathrm{x}_k, x_{k+1}, x_{k+2}, \ldots, x_m\right) \rightarrow \max \tag{3.10}$$

When searching for infimum, find:

$$f\left(\mathrm{x}_1, \mathrm{x}_2, \ldots, \mathrm{x}_k, x_{k+1}, x_{k+2}, \ldots, x_m\right) \rightarrow \min \tag{3.11}$$

subject to the following constraints:

$$\inf\left(X_i^\alpha\right) \leqslant x_i \leqslant \sup\left(X_i^\alpha\right), \quad i = k+1, k+2, \ldots, m, \tag{3.12}$$

$$\mathrm{x}_i \geqslant \inf(a_1^{iz})x_z + \inf(a_2^{iz}) \text{ for } i, z \in K;\ i \neq z;\ i \leqslant k,\ z > k,\ K \in K_s, \tag{3.13}$$

$$\mathrm{x}_i \leqslant \sup(a_1^{iz})x_z + \sup(a_2^{iz}) \text{ for } i, z \in K;\ i \neq z;\ i \leqslant k,\ z > k,\ K \in K_s, \tag{3.14}$$

$$x_i \geqslant \inf(a_1^{iz})x_z + \inf(a_2^{iz}) \text{ for } i, z \in K;\ i \neq z;\ i > k,\ z > k,\ K \in K_s, \tag{3.15}$$

$$x_i \leqslant \sup(a_1^{iz})x_z + \sup(a_2^{iz}) \text{ for } i, z \in K;\ i \neq z;\ i > k,\ z > k,\ K \in K_s, \tag{3.16}$$

The values a_1^{iz}, a_2^{iz} are the coefficients of interval regression equations determining a relation between variables X_i and X_z. Drawing values $(\mathrm{x}_1, \mathrm{x}_2, \ldots, \mathrm{x}_k)$ and determining $\hat{f}(\mathrm{x}_1, \mathrm{x}_2, \ldots, \mathrm{x}_k, \tilde{X}_{k+1}, \tilde{X}_{k+2}, \ldots, \tilde{X}_m)$ is repeated $\ddot{n}$ times. The overall procedure yields $\ddot{n}$ fuzzy sets characterised by membership functions $(\mu_1^f, \ldots, \mu_{\ddot{n}}^f)$. Thus, the value $\hat{f}(\hat{X})$ is represented by a fuzzy random variable.

Algorithm 1 Hybrid Propagation Method

Input: Function f, vector of correlated variables $\hat{X} = (X_1,\ X_2, \ldots, X_M)$
Output: Random fuzzy set defining $f(\hat{X})$

1: Define α_0, $\wp$, $\hat{n} = 1$
2: Generate a vector of correlated random values
   $(\mathrm{x}_1, \mathrm{x}_2, \ldots, \mathrm{x}_k)$
3: $\alpha = \alpha_0$
4: Determine α-levels $\tilde{X}_i^\alpha$, $i = k+1, k+2, \ldots, m$
5: Compute infimum (inf) and supremum (sup) of α-levels of a fuzzy number defining $\hat{f}(\hat{X})$ by solving the following two optimisation problems:
   $f\left(\mathrm{x}_1, \mathrm{x}_2, \ldots, \mathrm{x}_k, x_{k+1}, x_{k+2}, \ldots, x_m\right) \rightarrow \max$
   and $f\left(\mathrm{x}_1, \mathrm{x}_2, \ldots, \mathrm{x}_k, x_{k+1}, x_{k+2}, \ldots, x_m\right) \rightarrow \min$
   under the problem constraints specified by inequalities (3.12)–(3.16)
6: $\alpha = \alpha + \wp$
7: **If** $\alpha \leqslant 1$ **Then Go To** Step 4
8: $n = n + 1$
9: **If** $n \leqslant \ddot{n}$ **Then Go To** Step 2
10: **return** the set of fuzzy numbers $(\mu_1^f, \ldots, \mu_{\ddot{n}}^f)$

References

1. Fèron, R. 1976. Ensembles aléatoires flous. *Comptes Rendus de l'Académie des Sciences, Série A (Sciences mathématiques)* 282: 903–906.
2. Nahmias, S. 1978. Fuzzy variables. *Fuzzy Sets and Systems* 1(2): 97–110.
3. Stein, W.E., and K. Talati. 1978. Convex fuzzy random variables. *Fuzzy Sets and Systems* 6(3): 271–283.
4. Kwakernaak, H. 1978. Fuzzy random variables - I. definitions and theorems. Information. *Science* 15(1): 1–29.
5. Kwakernaak, H. 1979. Fuzzy random variables - II. algorithms and examples for the discrete case. Information. *Science* 17(3): 253–278.
6. Puri, M.L., and D. Ralescu. 1986. Fuzzy random variables. *Journal of Mathematical Analysis and Applications* 114(2): 409–422.
7. Diamond, P., and P. Kloeden. 1994. Metric spaces of fuzzy sets: Theory and applications. World Scientific.
8. Kruse, R. 1982. The strong law of large numbers for fuzzy random variables. *Information Science* 28(3): 233–241.
9. Krätschmer, V. 2001. A unified approach to fuzzy random variables. *Fuzzy Sets and Systems* 123(1): 1–9.
10. Liu, Y.-K., and B. Liu. 2003. Fuzzy random variables: A scalar expected value operator. *Fuzzy Optimization and Decision Making* 2(2): 143–160.
11. Liu, Y.-K., and B. Liu. 2003. Expected value operator of random fuzzy variable and random fuzzy expected value models. *International Journal of Uncertainty, Fuzziness and Knowledge-Based Systems* 11(2):195–215.
12. Liu, Y.-K., and B. Liu. 2002. Expected value of fuzzy variable and fuzzy expected value models. *IEEE Transactions on Fuzzy Systems* 10(4): 445–450.
13. Shafer, G. 1976. *A Mathematical Theory of Evidence*. Princeton: Princeton University Press.
14. Dempster, A.P. 1967. Upper and lower probabilities induced by a multivalued mapping. *The Annals of Mathematical Statistics* 38(2): 325–339.
15. Baudrit, C., D. Dubois, and D. Guyonet. 2006. Joint propagation and exploitation of probabilistic and possibilistic information in risk assessment. *IEEE Transaction on Fuzzy Systems* 14(5): 593–607.

Chapter 4
Multi-attribute Decision Making Process and Its Application

Abstract This chapter proposes the integrated fuzzy approach to solve Multi Attribute Decision Problems. Fuzzy Analytical Hierarchy Process (FAHP) is used to assign relative weights to criteria, and Technique for Order of Preference by Similarity to Ideal Solution (TOPSIS) is employed to rank the alternatives. The use of the proposed approach is illustrated using a real case from a steel industry.

Decisions regarding implementation of investment projects are crucial to the growth and success of each company. The quality of these decisions impacts the long-term effectiveness and market position of a company. However, financial and material limitations cause that not all potential investment projects can be implemented. So, almost every company sooner or later faces the problem of selecting a *portfolio of investment projects*. This problem is particularly important nowadays, in the age of omnipresent uncertainty in every business activity. Factors such as the increase in variability of products prices, growing pressure on reduction of the cost of production, increase of competitions, which can be observed in all branches of the economy, cause that the choice of an effective portfolio becomes an increasingly complex decision task. This motivates managers and decision makers (DMs) to use modern techniques and tools for managing the allocation of a company's capital, which in turn causes that methods for selecting an effective portfolio continue to be an important research area in the field of project management.

Selection of an effective portfolio should take into account not only financial aspects but also environment, market, organisation, technology, human resources and compliance with company's objectives. This leads to a multi-attribute evaluation of an investment. The main advantage of these approach is the ability to cover not just the financial aspects of the attractiveness of the investment project. It is quite obvious that the more parameters of investment projects subjected to analysis, the more reliable the results of the selection process of the projects. Unfortunately, the wider spectrum of criteria the higher the cost of analysis. Too many criteria may also cause that interpretation of results of analysis is very difficult.

Multi-attribute evaluation of investments belongs to a wider class of problems called Mutli-Attribute Decision Making (MADM) problems. A multi-attribute decision making (MADM) problem can be defined as:

© Springer International Publishing Switzerland 2015
I. Skalna et al., *Advances in Fuzzy Decision Making*,
Studies in Fuzziness and Soft Computing 333,
DOI 10.1007/978-3-319-26494-3_4

$$\begin{array}{c|ccc} & C_1 & \dots & C_n \\ \hline A_1 & f_{11} & \dots & f_{1n} \\ \vdots & \vdots & & \vdots \\ A_m & f_{m1} & \dots & f_{mn} \end{array}$$

$$w = [w_1, \dots, w_n]$$

where $A_1, \dots, A_n$ are possible alternatives, $C_1, \dots, C_m$ are criteria, f_{ij} is the rating of the alternative A_i with respect to the criterion C_j. The vector w is vector of weights, where w_j denotes the weight of the criterion C_j [1]. The goal of the MADM is to choose the best set of alternatives according to the defined criteria.

The MADM analysis requires that the selection be made among a predetermined, limited number of decision alternatives described by multiple, and often conflicting, criteria. The criteria are assigned weights relative to their importance, which then help a decision maker to make possibly the best decision. So, from the decision maker point of view, MADM analysis consists of methods that allow to combine information coming from different sources with additional information coming from a decision maker (e.g., preference) to rank available alternatives. The most important features of the MADM analysis are the following:

- Limited number of decision alternatives,
- Finite set of attributes describing each alternative.
- Discrete number of preference points.

Despite its undeniable usefulness, there are many problems related to classical MADM analysis, such as:

- Difficulties in the formulation of decision criteria.
- Competitiveness and inconsistency between criteria.
- Difficulties in the formulation of synthetic criterion based on sub-criteria.
- The presence of both quantitative and qualitative criteria.
- The uncertainty in parameter description.

The two latter problems are especially important. Usually, the values of decision criteria cannot be evaluated exactly. Moreover, the values of qualitative criteria estimated by different experts are difficult to compare.

The most common problems of the MAMD evaluation of projects in the presence of uncertainty are the following:

- How to take into account uncertainty during the process of evaluation of the value of criteria for each alternatives.
- How to establish appropriate fuzzy metrics for each qualitative criterion.
- How to appropriately propagate uncertainty through the multi-attribute decision making process.

4.1 General Steps of a Multi-attribute Decision Making Process

A typical multi-attribute decision making problem consists of the following three phases: identification, choice, ranking.

In the **identification phase**, a decision maker defines the decision context, i.e., *alternatives* and *decision criteria.* Alternatives represent available decisions or projects, whereas decision criteria measure how well each alternatives meet objectives. Each criterion must be measurable, or at least it must be possible to assess how much an alternative meets a criterion. So, an appropriate evaluation scale must be defined for each criterion. In this phase, projects are selected on the basis of the threshold criteria, which are determined by decision-makers. The value of each criterion for each alternative must strictly meet threshold criteria. Typical threshold criteria include financial criteria (e.g., $NPV > 0$, $IRR >$ threshold(IRR)), risk and strategy. Only those projects that meet threshold criteria pass to the next stage of the analysis.

In the **choice phase** all criteria are first arranged in a hierarchical order. Then, relative weights are assigned to all criteria. When the number of criteria is small enough, the weights can be assigned by experts. However, as the number of criteria included in evaluation process increases, the assignment of weights becomes a difficult task. In such cases, usually the *Analytical Hierarchy Process* (AHP) is used. In AHP, decision makers systematically compare criteria, two at a time, and determine which criterion is more important using some comparison scale (see Table 4.1). The comparison continues until all criteria have been compared. The comparison results are stored in the form of a matrix, which is called a *pairwise comparison matrix.* Then, the following steps are performed:

1. Calculate a priority vector in order to weight the elements of the comparison matrix.
2. Calculate global priorities by aggregating all local priorities using a simple weighted sum.
3. Use the eigenvalue in order to assess the strength of the consistency ratio of the comparative matrix and determine whether to accept the information. If a comparison matrix is not consistent, its elements should be adjusted and a consistency test should be carried out until the consistency ratio attains an acceptable value.

Table 4.1 Likert scale for a pairwise comparison

Definition	Intensity of importance	Fuzzy intensity of importance
Equally important	1	1
Moderately more important	3	(2, 3, 4)
Strongly more important	5	(4, 5, 6)
Very strongly more important	7	(6, 7, 8)
Extremely more important	9	(8, 9, 10)
Intermediate values	2, 4, 6, 8	

In the **ranking phase**, synthetic values of importance are computed for each alternative. There are many methods to obtain those values. Probably, the most popular is the *Technique for Order of Preference by Similarity to Ideal Solution* (TOPSIS) method, which is based on the assumption that the chosen alternative should have the shortest distance from the positive-ideal solution and the longest distance from the negative-ideal solution, where the positive ideal solution is the solution that maximises the benefit criteria and minimises the cost criteria, whereas the negative ideal solution is the solution that maximises the cost criteria and minimises the benefit criteria.

4.2 Uncertainty in MADM

Application of the approach presented above to real-life problems faces some difficulties. These are among others:

- The presence of quantitative and qualitative criteria.
- Description of some of evaluations using linguistic variables.
- Problems with comparison of criteria.
- Multiple and contradictory goals.
- Dependent projects.
- Uncertainty in data regarding specific criteria.
- A large number of feasible projects.
- Organisational requirements and controls.

To deal with those difficulties, the fuzzy approach is usually used instead of the crisp one. Probably the most important reasons for using the fuzzy approach to represent criteria and values of alternatives in MADM is the consistency with available information. In most problems of multi-attribute evaluation of investment projects, the subjective opinions on selected parameters are usually used. Neglecting imprecision and subjectivity of decision makers in pairwise comparison matrix may lead to errors in AHP methods.

The second problem is a determination of future values of the parameters determining the profitability of investment projects. However owing to the availability and uncertainty of information, it is very difficult to obtain the exact assessment data such as investment cost, gross income, expenses, depreciation, salvage value, interest rate, flexibility, productivity, quality etc. All of these factors, no matter tangible or intangible, are generally difficult to be quantified. In this situation most of decision-maker tend to give assessment based on their knowledge, past experience and subjective judgements. Linguistic terms such as “around 10 %”, “approximately between $300 000 and 450 000”, “about $80 000”, “very low”, “medium”, “high”, a frequently used to convey their estimations [2]. To deal with the vagueness of human thought fuzzy set theory can play significant role in this kind of decision making environment [3].

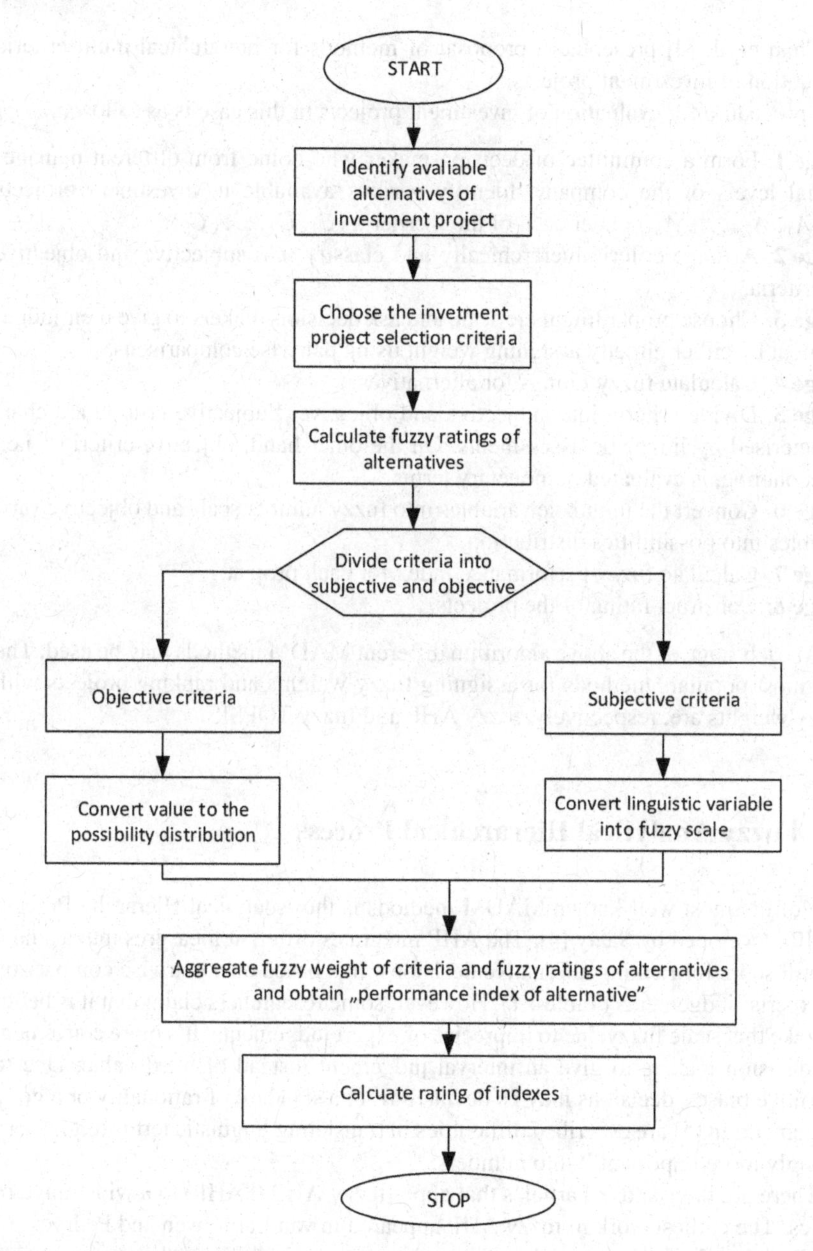

Fig. 4.1 The schematic diagram of investment selection

Hence a fuzzy multi-attribute decision making method is used to integrate various linguistic assessments and weights to determine the best investment project selection. In Fig. 4.1 method of fuzzy multi-attribute evaluation of investment is presented.

Chan et al. [3] presented a proposal of methods for hierarchical multi-criteria evaluation of investment projects.
The procedure for evaluation of investment projects in this case is as follows:

Stage 1. Form a committee of decision-maker who come from different managerial levels of the company. Identify various available m investment projects $(A_1, A_2, \ldots, A_m)$ under each of the k criteria $(C_1, C_2, \ldots, C_k)$,
Stage 2. Arrange criteria hierarchically and classify into subjective and objective criteria,
Stage 3. Choose proper linguistic scale and ask decision-makers to give their judgement by either directly assigning weight using pairwise comparisons,
Stage 4. Calculate fuzzy ratings for alternatives.
Stage 5. Divide criteria into subjective and objective. Subjective criteria are characterised by linguistic assessments. On the other hand, objective criterion, i.e., economic, is evaluated in monetary terms.
Stage 6. Convert the linguistic variables into fuzzy number scale and objective variables into possibilities distribution.
Stage 7. Calculate fuzzy performance index for each project.
Stage 8. Construct rating of the projects.

At each stage of the above algorithm different MADM methods may be used. The two most populars methods for assigning fuzzy weights and ranking projects with fuzzy weights are, respectively, fuzzy AHP and fuzzy TOPSIS.

4.3 Fuzzy Analytical Hierarchical Process

One of the most well-known MADM methods is the Analytical Hierarchy Process (AHP) developed by Saaty [4]. The AHP integrates different measures into a single overall score for ranking alternative decisions. It is based on a pairwise comparison of experts' judgements (Table 4.1). However, some researchers claim that it is better to make that scale fuzzy due to imprecise of expert judgements. It's more convenient for decision makers to give an interval judgement instead of fixed value. Due to cognitive biases, decisions may be deviated from a standard of rationality or a good judgement. In [5] are described difficulties in translating linguistic terms (e.g., "very strongly more important") into numbers.

There are thousands of articles that apply fuzzy AHP (FAHP) to a wide range of topics. The earliest work in fuzzy AHP appeared in van Larhooven and Pedrycz [6] and Buckley [7]. In these articles, comparison ratios are described by triangular and trapezoidal fuzzy numbers respectively. Chang [8] introduced a new approach for handling fuzzy AHP with use of the extended analysis method for making pairwise comparison. Kahraman et al. [9] use a fuzzy objective and subjective method obtaining the weights from AHP and make a fuzzy weighted evaluation. In 00's occurs thousands of articles about fuzzy numbers. Most important articles on theories and application of fuzzy AHP has more than 1000 citations [6–8]. Thus, fuzzy AHP can be considered as a well-established theory.

Algorithm 3 Buckley's Fuzzy AHP

1: Decision maker build the pairwise comparison matrix. Each element of comparison matrix is a trapezoidal fuzzy number $\tilde{t}_{ij} = (a, b, c, d)$,
2: For each row of pairwise comparison matrix the g index is calculated as a geometric mean of the row: $g_i = \left(\prod_{j=1}^{n} \tilde{t}_{ij}\right)^{1/n}$
3: The weight of i criterion is obtained using arithmetic mean of g_i

$$w_i = \frac{g_i}{\sum_{i=1}^{n} g_i}$$

Below, the most popular fuzzy approaches to AHP (Chang [8] and Buckley [7]) are presented. The Buckley's approach is as follow:

Algorithm 4 Chang's Fuzzy AHP approach

Input: The pairwise comparison matrix $T = [t_{ij}]$ consists of I rows and I columns. Each row and column represents a single criterion. The matrix is built by a decision maker.
1: Convert each element of the comparison matrix into a triangular fuzzy number $\tilde{t}_{ij} = (a, b, c)$.
2: Calculate fuzzy synthetic value:

$$w_j = \frac{\sum_{i=1}^{n} \tilde{t}_{ij}}{\sum_{i=1}^{n} \sum_{i=1}^{n} \tilde{t}_{ij}}.$$

3: For each pair of synthetic values calculate its degree of possibility:

$$V(w_i \geq w_k) = \sup(\min(\mu_{w_i}(x), \mu_{w_k}(y))).$$

For triangular fuzzy numbers this expression is equivalent to

$$V(w_i \geq w_k) = \begin{cases} 1, & \text{if } b_i > b_k, \\ 0, & \text{if } c_k \geq a_i, \\ \frac{a_i - u_k}{(b_k - c_k) - (b_i - a_i)}, & \text{otherwise.} \end{cases}$$

4: Weight of criterion i

$$a_i = \min_k (V(w_i \geq w_k))$$

5: Normalise the weights of criteria. AHP impact score of each criterion is measured by w index.

Table 4.2 Crisp evaluation matrix

	Coffee	Wine	Tea	Beer	Sodas	Milk	Water
Coffee	1	9	5	2	1	1	0.5
Wine	0.11	1	0.5	0.11	0.11	0.11	0.11
Tea	0.2	2	1	0.33	0.25	0.33	0.11
Beer	0.5	9	3	1	0.5	1	0.5
Sodas	1	9	4	2	1	2	0.5
Milk	1	9	3	1	0.5	1	0.33
Water	2	9	9	3	2	3	1

4.3.1 Example I—Comparison of AHP Methods

An exemplary problem of relative consumption of drinks in the United States, which was considered in [10], will be used to compare fuzzy and crisp AHP methods. Tables 4.2 and 4.3 present crisp and fuzzy evaluation matrix, respectively.

The above fuzzy evaluation matrix was obtained using pairwise comparison scale from Table 4.1. The weights of criteria was obtained using classical crisp algorithm, Buckley algorithms and Chang approach respectively. Buckley fuzzy weight are also defuzzified using formula:

$$w_j = \frac{a_j + b_j + c_j + d_j}{4} \tag{4.1}$$

The comparison is presented in Table 4.4. In column "real consumption" the actual consumption (from Statistical Abstract of the United States) are presented as a reference point.

The most important observations are as follows:

- The smallest difference is between real consumption and crisp approach, although it doesn't mean that crisp methods are the best. Fuzzy weight obtained from Buckley's approach represents also ignorance.
- Chang approach flatten weights of criteria.

4.3.2 Fuzzy TOPSIS

The second important problem in project portfolio selection is ranking of the projects. One of the most popular methods is fuzzy TOPSIS. The Technique for Order of Preference by Similarity to Ideal Solution (TOPSIS) is a multi-criteria decision analysis method. It was developed by Hwang and Yoon [11] with further developments by Yoon [12] and Hwang et al. [13].

Like it was shown during AHP presentation, there are many approaches to Fuzzy TOPSIS methods. Below, the method proposed by Chen and Hwang [14] is presented.

Table 4.3 Fuzzy evaluation matrix

	Coffee	Wine	Tea	Beer	Sodas	Milk	Water
Coffee	(1, 1, 1, 1)	(8, 9, 9, 10)	(4, 5, 5, 6)	(1, 2, 2, 3)	(1, 1, 1, 1)	(1, 1, 1, 1)	(0.33, 0.5, 0.5, 1)
Wine	(0.1, 0.11, 0.11, 0.125)	(1, 1, 1, 1)	(0.33, 0.5, 0.5, 1)	(0.1, 0.11, 0.11, 0.125)	(0.1, 0.11, 0.11, 0.125)	(0.1, 0.11, 0.11, 0.125)	
Tea	(0.167, 0.2, 0.2, 0.25)	(1, 2, 2, 3)	(1, 1, 1, 1)	(0.25, 0.33, 0.33, 0.5)	(0.2, 0.25, 0.25, 0.33)	(0.25, 0.33, 0.33, 0.5)	(0.1, 0.11, 0.11, 0.125)
Beer	(0.33, 0.5, 0.5, 1)	(8, 9, 9, 10)	(2, 3, 3, 4)	(1, 1, 1, 1)	(0.33, 0.5, 0.5, 1)	(1, 1, 1, 1)	(0.33, 0.5, 0.5, 1)
Sodas	(1, 1, 1, 1)	(8, 9, 9, 10)	(3, 4, 4, 5)	(1, 2, 2, 3)	(1, 1, 1, 1)	(1, 2, 2, 3)	(0.33, 0.5, 0.5, 1)
Milk	(1, 1, 1, 1)	(8, 9, 9, 10)	(2, 3, 3, 4)	(1, 1, 1, 1)	(0.33, 0.5, 0.5, 1)	(1, 1, 1, 1)	(0.33, 0.5, 0.5, 1)
Water	(1, 2, 2, 3)	(8, 9, 9, 10)	(8, 9, 9, 10)	(2, 3, 3, 4)	(1, 2, 2, 3)	(2, 3, 3, 4)	(1, 1, 1, 1)

Algorithm 5 Fuzzy TOPSIS

Input: The input variable are decision matrix D and weights of criteria. Each row of D represent j-th alternative and each column i-th criterion.

$$D = \begin{bmatrix} x_{11} & \cdots & \\ \vdots & \ddots & \vdots \\ & \cdots & x_{mn} \end{bmatrix}$$

The value of x_{ij} is represented by fuzzy number.

1: **Normalisation using linear scale.** Each element of the matrix D is transformed using linear scale:

$$D_{norm} = \left(r_{ij}\right)_{mxn} = \frac{x_{ij}}{\sum_{i=1...m} x_{ij}}$$

The normalisation is performed in order to eliminate anomalies with different measurement units and scales.

2: **Calculation of normalised decision matrix**. For each element of D_{norm} calculate the weighted normalised decision matrix

$$\left(v_{ij}\right)_{mxn} = \left(w_j r_{ij}\right)_{mxn},$$

where w_j is a weight of j-criterion obtained from FAHP methods.

3: **Positive Ideal Solution/Negative Ideal Solution**: Determine Positive Ideal Solution (PIS) and Negative Ideal Solution (NIS). PIS and NIS are defined as

$$a_{PIS} = [v_1^{max}, ..., v_n^{max}]$$

$$a_{NIS} = [v_1^{min}, \ldots, v_n^{min}]$$

where $v_j^{max} = \max_i v_{ij}$ and

$$v_j^{min} = \min_i v_{ij}$$

To obtain maximal value of fuzzy data, some ranking procedures should been made. This example shows different methods of ranking fuzzy numbers approaches.

4: **Separation Measure** - Obtain Separation Measures from PIS and NIS. Classically, this is a Manhattan or Euclidean measure. In order to simply the approach the Manhattan Measure is used

$$S_i^{PIS} = \sum_{i=1}^{n} \left| v_{ij} - v_j^{max} \right|$$

$$S_i^{NIS} = \sum_{i=1}^{n} \left| v_{ij} - v_j^{min} \right|$$

Difference between fuzzy numbers are described in Chap. 1.

5: **Closeness** – Compute Relative Closeness to Ideal

$$C_i = \frac{S_i^{NIS}}{S_i^{PIS} + S_i^{NIS}}$$

Table 4.4 Importance of the drinks: fuzzy weights, crisp weights and real consumption

	Buckley fuzzy weights	Buckley after deffuzification	Chang approach	Crisp	Real consumption
Water	(0.18, 0.325, 0.325, 0.546)	0.344	0.200	0.323	0.240
Sodas	(0.109, 0.189, 0.189, 0.332)	0.205	0.000	0.188	0.267
Coffee	(0.114, 0.178, 0.178, 0.294)	0.191	0.000	0.177	0.133
Milk	(0.089, 0.13, 0.13, 0.204)	0.138	0.835	0.129	0.129
Beer	(0.076, 0.118, 0.118, 0.204)	0.129	0.235	0.125	0.173
Tea	(0.024, 0.039, 0.039, 0.067)	0.042	0.072	0.039	0.040
Wine	(0.014, 0.021, 0.021, 0.032)	0.022	0.410	0.020	0.014

As a result a vector C is obtained. The elements of the vector C represent the indexes for each alternative. In the next step alternatives are ranked in descending order. This is the final ranking for projects.

4.4 Example—Investment Strategy of the Steel Company

A board of directors wants to develop investment strategy of a steel company. They take into consideration 10 potential investment and 23 attributes.

4.4.1 Calculation of Fuzzy Ratings

To solve such problem the analyst should construct hierarchy of the criteria. The tree of the criteria is presented on Fig. 4.2. Tree consists of three layers—criteria, sub-criteria and attribute. Factors belong to criteria layer are marked by Cx, sub-criteria layer $Cx.x$, and attribute layer $Cx.x.x$. There are five main group of criteria (first layers). There are five criteria—financial, market, technology and environment, staff and compliance with the company's strategic objective. Each of them is divided into subcriteria. The following objective subcriteria were used:

- **NPV**—is defined as a sum of discounted net cash flows appearing in consecutive years of the economic life of an investment project.
- **IRR**—is the annualised effective compounded return rate that makes the NPV from a particular investment equal to zero. It can also be defined as the discount rate at which the present value of all future cash flow is equal to the initial investment or in other words the rate at which an investment breaks even.
- **Payback period**—refers to the period of time required to recoup the funds expended in an investment.

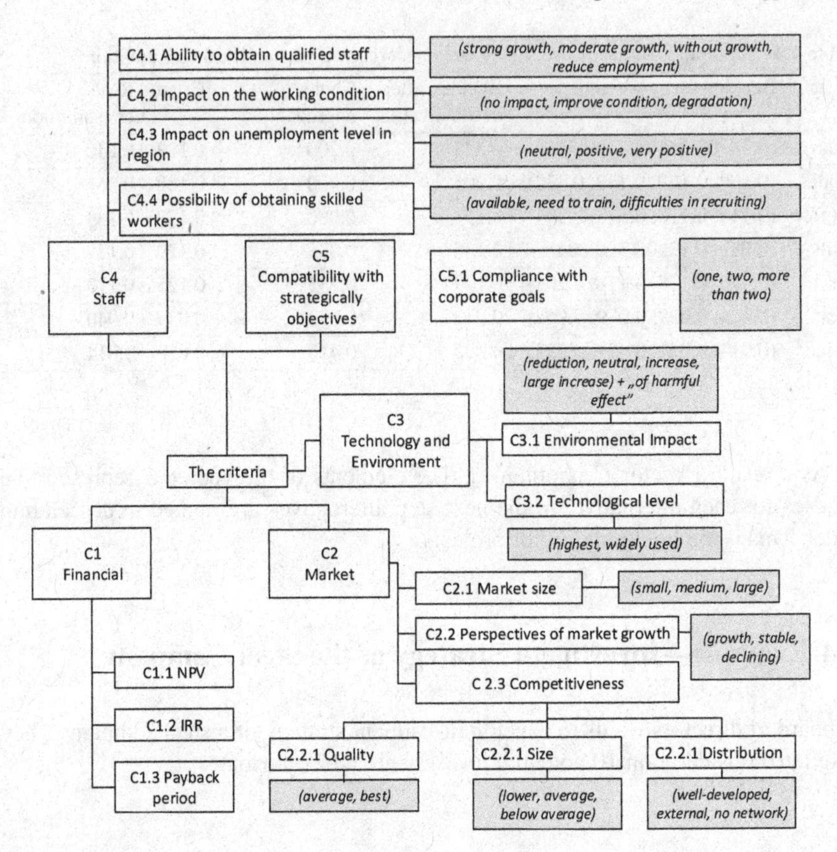

Fig. 4.2 Hierarchy of factors

Table 4.5 Pair-wise comparison matrix of criteria (level 1)

		C1	C2	C3	C4	C5
Financial criteria	C1	1	5	9	9	9
Market	C2	0.2	1	3	3	3
Technological and Environmental	C3	0.11	0.33	1	1	1
Staff	C4	0.11	0.33	1	1	1
Compatibility with the strategical objective	C5	0.11	0.33	1	1	1

After determination of criteria hierarchy, the pairwise comparison matrix for each layer was prepared. The exemplary matrix for first level of criteria is presented in Table 4.5.

In this example six pairwise comparison matrix were prepared—one for criteria level, five for sub criteria level (with exception for C5.x), and one for attribute level.

Table 4.6 Summary of consistency check

	Consistency ratio
Criteria	0.04
Subcriteria C1.x	0
Subcriteria C2.x	0.07
Subcriteria C3.x	N/A
Subcriteria C4.x	0.05
Attributes	0.03

Simultaneously, definitions of scales for each decision factors were prepared. There are two type of scales—numerical and linguistic. Only three factors—NPV, IRR and Payback period was described by fuzzy numbers, the rest of subcriteria was subjective. The value of linguistic attributes are presented as a leaf in grey boxes, the rest of factors were numerical and described by fuzzy numbers.

Building investment strategy has started from calculation of weight factors using AHP method. Each expert prepared pairwise comparison matrices in terms of fuzzy numbers. Then the consistency on each of the matrices was verified to ensure that all consistency requirements was satisfied. Consistency ratio of each matrix is presented in Table 4.6.

Next, all comparisons are converted into fuzzy scale and weights are calculated using Buckley's approach. Values of the weight of criteria are presented in Table 4.7.

Table 4.7 Importance weights of individual requirements

Weight		Weight	
C1	(0.429, 0.633, 0.917)	C3.1	(0.039, 0.056, 0.124)
C2	(0.073, 0.175, 0.372)	C3.2	(0.006, 0.008, 0.018)
C3	(0.051, 0.064, 0.122)	C4.1	(0.003, 0.007, 0.039)
C4	(0.051, 0.064, 0.122)	C4.2	(0.013, 0.036, 0.145)
C5	(0.051, 0.064, 0.122)	C4.3	(0.008, 0.019, 0.085)
C1.1	(0.181, 0.292, 0.471)	C4.4	(0.001, 0.003, 0.013)
C1.2	(0.181, 0.292, 0.471)	C5.1	(0.051, 0.064, 0.122)
C1.3	(0.025, 0.042, 0.072)	C2.3.1	(0.002, 0.027, 0.313)
C2.1	(0.006, 0.05, 0.307)	C2.3.2	(0.001, 0.02, 0.264)
C2.2	(0.018, 0.061, 0.204)	C2.3.3	(0.002, 0.015, 0.127)
C2.3	(0.011, 0.063, 0.31)		

4.4.2 Computation of Performance Index of Criteria

To compute performance index the fuzzy evaluation matrix was created. Boards of directors took into consideration 10 alternatives:

- **P1**—Modernisation of the heavy section mill,
- **P2**—Increase of the capacity of the hot rolling mill,
- **P3**—Construction of the cold rolling mill with the capacity of 1000 thousand t/year,
- **P4**—Construction of the cold rolling mill with the capacity of 1500 thousand t/year,
- **P5**—Construction of the hot dip galvanising line with the capacity: 300 thousand t/year,
- **P6**—Construction of the hot dip galvanising line with the capacity: 400 thousand t/year,
- **P7**—Construction of the organic coating line with the capacity of 200 thousand t/year,
- **P8**—Construction of the organic coating line with the capacity of 300 thousand t/year,
- **P9**—Construction of the tinning plant with the capacity of 100 thousand t/year,
- **P10**—Construction of the wire drawing plant.

For each investment project the objective criteria were characterised by fuzzy numbers which resulted from the hybrid simulation described in Sect. 1.3.1. The levels of subjective criteria are specified by experts. Values of subjective criteria were translated into fuzzy numbers using procedure as follows. In the presented example, there are two kinds of subjective attributes—some of them describe patterns, and some of them judgements. For example, *market size* criterion C2.1 and *prospects for market growth* criterion C2.2 belong to the first group. They describe the belief of decision maker that market for alternatives will behave in accordance with some pattern. For example, pattern had been stable means the dynamic of the market growth which may be described by the fuzzy number $(-1.02, 0, 1.02)$. The second group that is subjective criteria represents judgements of experts. Therefore, they are treated as ordinal fuzzy variables. Description of project alternatives is presented in Fig. 4.3.

Exemplary translation of subjective criteria for one factor is presented in Table 4.8 and coding of one alternative is presented in Table 4.9.

The translation table is used to create the so-called *performance matrix*. Rows in the latter matrix describe alternatives and column correspond to criteria. The entries, described by fuzzy numbers, asses how well each option performs with respect to each of the criteria. Then, using FTOPSIS, the ranking of criteria was constructed (Table 4.10).

Projects	Capital investment	C1			C2					C3		C4				C5
	000's PLN	C1.1	C1.2	C1.3	C2.1	C2.2	C2.3			C3.4	C3.5	C4.1	C4.2	C4.3	C4.4	C5.1
							C2.3.1	C.2.3.2	C.2.3.3							
P1	(135,150,165)	(-66 083.9, 14 376.1, 127 533.0,225 122.1)	(-5.1,4.5, 9.3,12.1)	2.1	middle	stability	average	averange	external	widely used	neutral	without growth	no impact	positive	avaliable	one
P2	(216,240,264)	(-1 334 440.0, 628 247, 493 130.0, 1 757 191.0)	(-4.3,-1.0, 3.3,7.4)	4.5	big	stability	average	averange	well-developed	widely used	increase	moderate growth	no impact	nautral	need to train	one
P3	(1 270,1 430,1 590)	(-1 00 0694.0, -24 717, 659 279.1, 1779 179.1)	(-2.3,-0.9 , 2.3,5.9)	5.2	big	stability	average	averange	well-developed	widely used	increase	moderate growth	no impact	positive	need to train	more than two
P4	(1 600,1 780 ,1 995)	(-258 273.0, -111 259.0, 202 888.0, 5 023 315.0)	(-1.1,-0.8 , 1.3,6.2)	4.6	big	stability	average	averange	well-developed	widely used	large increase	moderate growth	no impact	positive	need to train	more than two
P5	(375, 410, 450)	(-356 417.0, -140 463.0, 291 199.4, 785 944.0)	(-3.3,-0.8 , 2.4, 7.9)	3.2	big	growth	best	below averange	well-developed	widely used	large increase	moderate growth	improve condition	neutral	need to train	more than two
P6	(465, 515, 565)	(-368 432.0, -102371.0, 560 595.0, 1 123 142.0)	(-3.0,-0.5 , 2.9, 7.8)	3.1	big	growth	average	averange	well-developed	widely used	increase	moderate growth	degradation	very positive	need to train	more than two
P7	(125, 138 , 150)	(-102 722, -50 513.0, 245 279.2, 391 245.1)	(-1.7 ,-0.8, 5.9, 9.1)	2.1	big	growth	best	averange	well-developed	highest	increase	moderate growth	degradation	very positive	need to train	more than two
P8	(170, 190, 210)	(-144 482, -70 168.0, 235 800.5, 448 462.5)	(-2.8, -0.6 , 4.9, 7.9)	2.3	big	growth	best	averange	well-developed	highest	increase	moderate growth	degradation	very positive	need to train	more than two
P9	(250, 320, 350)	(-137 252.0, 23 797.3, 129 599.0, 301 074.2)	(-3.8, 1.4 , 5.9, 6.9)	3.3	small	stability	best	below averange	no network	highest	large increase	moderate growth	degradation	positive	need to train	more than two
P10	(20, 22, 24)	(-12 344.0, 4 322.1, 16 123.0, 28 144.0))	(-4.1, 2.2, 5.9, 7.6)	4.1	big	small	average	lower	no network	widely used	neutral	moderate growth	no impact	neutral	difficulties in recruiting	one

Fig. 4.3 Description of the projects alternatives

Table 4.8 Coding of exemplary alternative

Projects		P1 value of attributes	P1 numerical code of attributes
Capital investment	000's PLN	(135, 150, 165)	(216, 240, 264)
C1	C1.1	(−66083.9, 14376.1, 127533.0, 225122.1)	(−66083.9, 14376.1, 127533.0, 225122.1)
	C1.2	(−5.1, 4.5, 9.3, 12.1)	(−5.1, 4.5, 9.3, 12.1)
	C1.3	2.1	2.1
C2	C2.1	Middle	(1100, 1300, 1400, 1600)
	C2.2	Stability	(−1.02, 1.0, 1.02)
	C2.3.1	Average	(0, 0.1, 1)
	C.2.3.2	Average	(0.95, 1, 1.05)
	C.2.3.3	External	(0, 0.5, 1)
C3	C3.1	Widely used	(0, 0.1, 0.2)
	C3.2	Neutral	(0, 0.5, 1)
C4	C4.1	Without growth	(0, 0.1, 1)
	C4.2	No impact	(0.5, 1, 1)
	C4.3	Positive	(0, 0, 0.5)
	C4.4	Available	(0, 0.33, 0.66)
C5	C5.1	One	(0, 0.2, 0.4)

Table 4.9 Exemplary transformation of subjective criteria

Linguistic value	Fuzzy value
Not important	(0, 0, 0.33)
Neutral	(0, 0.33, 0.66)
Important	(0.33, 0.66, 1)
Very important	(0.66, 1, 1)

Table 4.10 Final ranking of projects

Project	Rank
P9	0.7154
P10	0.7101
P1	0.7095
P4	0.7011
P3	0.6817
P7	0.6789
P8	0.6782
P5	0.6770
P6	0.6714
P2	0.6615

References

1. Yu, P.-L. 2013. *Multiple-criteria decision making: concepts, techniques, and extensions*, vol. 30. Springer Science and Business Media.
2. Ward T.L. 1985. Discounted fuzzy cash flow analysis. In *Fall industrial engineering*, 476–481.
3. Chan, F.T.S., M.H. Chan, and N.K.H. Tang. 2000. Evaluation methodologies for technology selection. *Journal of Materials Processing Technology* 107(1–3): 330–337.
4. Saaty, T.L. 1980. *The analytic hierarchy process: planning, priority setting resource allocation.*, Advanced Book Program New YorK: McGraw-Hill.
5. Bocklisch, F., S.F. Bocklisch, and J.F. Krems. 2010. *How to translate words into numbers? Fuzzy approach for the numerical translation of verbal probabilities*, vol. 6178., Lecture Notes in Computer Science Berlin: Springer.
6. van Laarhoven, P.J.M., and W. Pedrycz. 1983. A fuzzy extension of saaty's priority theory. *Fuzzy Sets and Systems* 11(1–3): 199–227.
7. Buckley, J.J. 1985. Fuzzy hierarchical analysis. *Fuzzy Sets and Systems* 17(3): 233–247.
8. Chang, D.-Y. 1996. Applications of the extent analysis method on fuzzy AHP. *European Journal of Operational Research* 95(3): 649–655.
9. Kahraman, C., D. Ruan, and E. Tolga. 2002. Capital budgeting techniques using discounted fuzzy versus probabilistic cash flows. *Information Science* 142(1–4): 57–76.
10. Saaty, T.L., and L.T. Tran. 2007. On the invalidity of fuzzifying numerical judgments in the analytic hierarchy process. *Mathematical and Computer Modelling* 46(7–8): 962–975.
11. Hwang, C.-L., and K. Yoon. 1981. *Multiple attribute decision making: methods and applications a state-of-the-art survey*, vol. 186., Lecture Notes in Economics and Mathematical Systems Berlin: Springer.
12. Yoon, K. 1987. A reconciliation among discrete compromise solutions. *The Journal of Operational Research Society* 38(3): 277–286.

13. Hwang, C.-L., Y.-J. Lai, and T.-Y. Liu. 1993. A new approach for multiple objective decision makings. *Computers and Operational Research* 20(8): 889–899.
14. Chen, S.-J.J., and C.L. Hwang. 1992. *Fuzzy multiple attribute decision making: methods and applications*, vol. 375., Lecture Notes in Economics and Mathematical Systems New York: Springer.

Chapter 5
Risk Assessment in the Presence of Uncertainty

Abstract This chapter is devoted to a method which is able to process hybrid data, i.e., to jointly handle both randomness and imprecision. Random variables are described by probability distributions and imprecise values are modelled using possibility distributions. The main advantage of the proposed method is that it takes into account the dependencies between economic parameters.

At present, methods for risk assessment constitute a fundamental tool supporting decision-making processes. The risk associated with decisions arises from the fact that economic parameters are usually burdened with uncertainty.

5.1 Description of Uncertainty in the Economic Risk Assessment

Risk and *uncertainty* are defined in various ways in the world literature. The definition that will be adopted in this book says that risk results from actions taken by humans and depends functionally on uncertainty, whereas uncertainty concerns objective states of nature, that is the inability to predict and accurately identify future business. The uncertainty can be connected with external (e.g., income tax rates, raw material and energy prices, forecasted market size, interest rates, exchange rates etc.) and internal (e.g., product prices, material consumption indicators, labour consumption indicators etc.) conditions of company's activity. Therefore, the economic risk is most frequently determined as a possibility of occurrence of unfavourable values of the measure of the efficiency of the business activity. It is also identified with variableness of the measure of the efficiency of the business activity [1].

Quantification of risk is one of the most difficult tasks in economic risk management. A chief problem in this phenomena is not only to develop methods for estimation of the economic risk, but also to improve methods for data gathering and processing for a formal description of uncertainty of parameters of the risk calculus. An adequate description of uncertainty of those parameters has a decisive meaning

© Springer International Publishing Switzerland 2015

I. Skalna et al., *Advances in Fuzzy Decision Making*,
Studies in Fuzziness and Soft Computing 333,
DOI 10.1007/978-3-319-26494-3_5

in estimation of the risk. It is a condition of effective application of quantification methods of economic risk assessment in practice.

For many years, the only tool that allowed to express uncertainty in mathematical language was probability calculus. In fact, it still remains the most common tool used in practice and prevails in the literature concerning the economic risk. The probabilistic approach to risk analysis employs stochastic simulation to evaluate the risk. However, high cost of data preparation and difficulties with determining probability distributions of economic parameters significantly limit the usage of this approach. Furthermore the last decades have shown that the number and complexity of dependencies both inside and outside a company makes it difficult to use the probability theory to represent all kinds of the uncertainty appearing in case of the economic risk assessment. That is why, apart from quantitative methods, qualitative methods are also utilised to predict the values of economic parameters and to assess the risk. Experts' opinions and subjective probability distributions are then used.

Qualitative methods are more important, since in numerous decision-making situations uncertainty of economic parameters is not probabilistic in nature, but rather results from insufficient or vague information and is epistemologically indeterminate [2]. Sometimes, as pointed by Gupta [3], it happens that uncertainty is probabilistic, but the available information is rather fuzzy. In practice, quite often it is not possible to determine probability distribution, because of no (and there is no option to get) sufficient volume of data facilitating execution of statistical tests. On the other hand, assumption of "no data available at all" is also not true. In general, there is always some information available. These could be estimates of unknown values made by experts. A good example is the problem of assessment of profitability and risk of project investments. The main difficulty in this problem stems from the uniqueness of each investment project and the time interval between the moment of studies on a project and its realisation and exploitation. Because of the uniqueness, it is usually difficult to predict all possible values that the parameters of efficiency calculus can take and to determine the probability of their realisation [4]. Pluta and Jajuga [5] say that when assessing the risks attached to investment projects, usually only experts' opinions and subjective probability distributions of the possible values of parameters can be used. The estimation of net present value (*NPV*) expected values from historical data is much more difficult because of the specificity and uniqueness of investment projects. Namely, it is not possible to obtain perfectly reliable information regarding similar past projects. Each company is unique in its business model, investment needs, production capabilities, liaisons with the environment. Each company goes through different stages of development during its lifecycle.

The estimate of subjective probability is based on the experience of a person which determines the probabilities of occurrence of individual events. These probabilities represent the level of conviction of an expert that the event will occur [4]. Dittmann [6] interprets the subjective probability assigned to experts' forecasts as a chance that the predicted value is equal approximately the actual value of the analysed variable. Subjective probability can also be interpreted as a possibility of occurrence of an event [4, 6]. The level of subjectivity depends on the way of estimation and the knowledge on other similar event. In practice, an arbitrary probability distribution between minimal and maximal estimation is usually adopted.

The usage of subjective probability distributions in the economic risk assessment causes many problems associated with estimation of those distributions. Choobineh and Behrens [1] and Kuchta [4, 7] point out these difficulties. Kuchta says that sometimes a decision-maker does not know how to answer the question on the probability of the unique, unrepeatable event. The question about frequency has not much sense. Decision-maker can have, however, a point of view on the degree of possibility of occurrence of respective values [4]. Moreover, the subjective probability distribution must have the same properties as any probability distribution. For example, the sum of probabilities of all elementary events must sum up to 1 and the probability of the simultaneous occurrences of two independent events is the product of the probabilities of each event. It is extremely difficult to maintain these properties in expert judgements about subjective probability of future values [4]. This problem can be partially solved by modelling uncertainties using fuzzy numbers. When uncertainty is described using fuzzy numbers, a decision-maker can give arbitrary values of possibility degrees according to own feelings. Fuzzy approach does not impose the form of expression of subjective opinions as much as probabilistic approach does [4]. Mohammed and McCowan [8] argue that for most practitioners triangular and trapezoidal fuzzy numbers are much easier to understand and to apply than probability distributions. People hardly think in probabilistic terms, fuzzy sets notation or linguistic description of uncertainty seems to be more natural and much closer to human thinking. The construction of a triangular fuzzy number based on the best, the worst and average values is closer to the possibility theory than to the probability theory [4]. Moreover, many authors questions the legitimacy of modelling the absolute lack of knowledge about selected parameter using uniform probability distribution [2, 9].

In most of the existing approaches, different ways of uncertainty representation are usually unified in a single modelling framework. In order to perform the unification, it is necessary to be able to transform one form of uncertainty into another. Obviously, such transformation is not without problems. For example, transformation of a probability distribution into a possibility distribution causes the loss of information, whereas the opposite one requires additional information to be introduced. This leads eventually to systematic errors in risk assessment, i.e., overestimation or underestimation of the risk, depending on the direction of transformation. The most appropriate approach to risk assessment is to develop and use methods which allow hybrid representation of uncertainty, i.e., expressed by probability distributions, fuzzy numbers to be processed according to their nature and only finally combine them into a synthetic easy-to-interpret risk measure. The research of the subject [2, 9, 10] shows that models with hybrid data can be successfully used to support decision-making in economic business.

Taking into consideration the above mentioned problems, most of the real-world evaluations of risk contain a mixture of quantitative and qualitative data. Two methods of description of the uncertainty of the economic calculus parameters (probability distribution, fuzzy numbers) are used usually as alternatives. The most common situation in practice is when for some parameters it is possible to determine probability distributions, while for some, information is available in form of fuzzy numbers. In case of economic calculus, data which is available, is usually heterogeneous, uncertain and imprecise, and it is usually coming from various sources. These are both statistical data as well as subjective assessments of phenomena made by experts [2, 3, 9–15].

Ward [16] was the first to utilise fuzzy numbers in financial analysis. Ward presented cash flows by means of trapezoidal fuzzy numbers. Buckley [17] uses fuzzy numbers for calculation of net present values of investment projects. Calzi [18] presented principles of expanding financial mathematics for fuzzy numbers. Choobineh and Behrens [1] present the use of possibility distributions in economic analysis. Chiu and Park [11] apply fuzzy numbers in the calculation of the efficiency of investment projects. The authors introduce the methods of the choice of one project, from the set of mutually exclusive projects. Esogbue and Hearnes [19] utilise fuzzy numbers in problems concerning the replacement of fixed assets. Their aim was to describe the economic life-cycle of fixed assets. Kahraman et al. [20] present methods calculating selected investment project effectiveness ratios on the assumption that certain parameters are presented in the form of fuzzy numbers. Kuchta presents the use of fuzzy numbers in capital budgeting [7]. In [13] the results of the evaluation of the profitability and the risk of investment projects in case when the uncertainty of parameters is presented in the form of probability distributions and fuzzy numbers are compared. Rębiasz [12, 21–23] presents the usage of fuzzy numbers and probability distribution for the evaluation of projects and selection of the most profitable project from the steel industry.

Opinions of authors on usefulness of fuzzy and probabilistic approach in decision-making analysis vary. Majority of authors argue that fuzzy and probabilistic approaches are supplementary to each other, and in each case it must be decided which approach will be more adequate. The selection of an appropriate approach should be conditioned mostly on the degree of subjectivity of the available information. On the other hand, Gupta [3] and Smets [24] claim that in decision-making, probabilistic description of uncertainty is more effective than description using possibility distribution. They say, moreover, that a decision-maker is not interested in "what is possible" but in "what is probable". Thus, they suggest purposefulness of transformation of possibility distribution into probability distribution. Kuchta [4], in turn, says that selection of the method of uncertainty representation depends mainly on the experience and habits of the decision-maker. Choobineh and Behrens [1] claim that maintaining the probabilistic approach stems more from tradition then from conscious selection.

In practice, the most common situation is when for some parameters it is possible to determine probability distributions, while for others information is available in form of fuzzy numbers. It must be underlined that probabilistic approach and fuzzy approach are not contradictory and can be applied simultaneously. The probability theory and the possibility theory emphasize different aspects of uncertainty. The probability theory offers quantitative model of randomness, whereas the possibility theory describes qualitative model of incomplete knowledge [2–4, 25, 26]. Baudrit et al. [2] argue that randomness and imprecise or missing information are two reasons of uncertainty, which have an impact on risk. Thus, it is necessary to include these two approaches to description of uncertainty in the process of risk assessment. The usage of both probability and possibility distributions allows to reflect more properly the knowledge on economic parameters. However, for the time being there are few studies which are devoted to the use of hybrid representation of uncertainty [2, 26]. As suggested by Ferson and Ginzburg [14] distinct methods are needed to adequately represent random variability, often referred to as "objective uncertainty", and imprecision often referred to as "subjective uncertainty". In risk assessment, no distinction is traditionally made between these two types of uncertainty, both being represented by means of a single probability distribution [2–4, 25, 26]. In case of partial ignorance, the use of a single probability measure introduces information that is in fact not available. This may seriously bias the outcome of risk analysis in a non-conservative manner. Kaufman and Gupta [27] introduced hybrid numbers, which simultaneously express inaccuracy and randomness. Guyonnet et al. [26] proposed a method of risk estimation in the case when both probability and possibility distributions are used to represent uncertainty. This method is a modification of the method previously developed by Cooper et al. [15]. It combines stochastic simulation with arithmetic of fuzzy numbers. The result of processing such data is given in the form of two cumulative distribution functions: optimistic and pessimistic. Similarly, Baudrit et al. [2] use probability and possibility distributions in risk analysis. Their procedure also combines stochastic simulation with arithmetic on fuzzy numbers, but they give the result in the form of a fuzzy random variable, which characterises the examined phenomenon.

The use of hybrid data causes many problems. They involve primarily the problem of dependency between model parameters. Economic problems often involve parameters that are mutually correlated. For example, there is a correlation between enterprise product prices and raw material prices, also the volumes of sales of different assortments are correlated. Usually, when processing such data, the independence of the parameters is assumed [2, 26]. However, the omission of this dependency leads to systematic errors in risk quantification, usually results in a large overestimation of the actual risk. Another problem connected with the use of hybrid data is the lack of universally accepted and easy interpretable measures that synthetically express the risk of an economic activity. So far in the literature there is no solution to these problems. Construction of methods which will be able to take into account dependencies (correlations) between uncertain parameters would enable the risk to be assessed far more accurately. Solution of the dependence problem can potentially broaden the scope of application of such methods to a wide range of decision problems.

The second weakness is the lack of synthetic risk measures (indexes) readable for decision-makers. A method for risk appraisal can be accepted by economic life practitioners only if it describes the risk in the form of synthetic indexes readable and comprehensible for them. In the papers mentioned above risk was expressed by optimistic and pessimistic cumulative distribution functions of the examined financial ratio. Such cumulative distributions are difficult to interpret and often unintelligible for applied decision-makers. Moreover, in the literature there is no consistent opinions on how to designate optimistic and pessimistic cumulative distribution [2, 12].

Several numerical and symbolic methods have been proposed for handling uncertain information. Three of the most common frameworks for representing and reasoning with uncertain knowledge are [1, 2, 4]:

- Bayesian probability theory.
- Dempster–Shafer (D–S) theory of evidence.
- Fuzzy set theory.

Each of these frameworks is aimed at a special application environment and has its own features.

5.2 Measures of Risks in the Case of Hybrid Data

In order to effectively use fuzzy random variables in risk assessment, it is necessary to define the expected value and the variance for these variables. Many authors define the expected value for such variables in various ways. Most frequently, it is defined in the form of a fuzzy set [28]. However, in decision-making problems, the expected value is desired in a scalar form [29–31]. This facilitates interpretation of results. Methods using such values are easily accepted by practitioners.

In case of the application of hybrid data the risk can be measured using standard semi-deviation of analysed indicator calculated on the basis the fuzzy number or the fuzzy random set or the upper and lower cumulative distribution function calculated on the basis fuzzy random set.

The concept of the expected value is introduced for fuzzy variables. Liu and Liu [29–31] define the expected value $E(\xi)$ of a fuzzy variable ξ as follows:

$$E(\xi) = \int_{-\infty}^{+\infty} x d\Phi(x), \tag{5.1}$$

where Φ is a credibility distribution, if conditions $\lim_{x \to -\infty} \Phi(x) = 0$ and $\lim_{x \to +\infty} \Phi(x) = 1$ are satisfied.

Standard deviation $SDev(\xi)$ of fuzzy variable ξ, which has a finite expected value e, is defined by:

$$SDev(\xi) = \sqrt{E((\xi - e)^2)}. \tag{5.2}$$

Liu and Liu [29–31] proposed a new method for calculating the expected value and the variance of a fuzzy random variable. The basic concept of these values is based on the credibility of a fuzzy event [29–31].

Definition 5.1 Let ξ be a normal random fuzzy variable defined on a possibility space $(\Theta, \mathcal{P}(\Theta), Pos)$. The expected value, $E(\xi)$, of ξ is defined by:

$$E(\xi) = \int_{\Omega} \left[\int_{0}^{\infty} Cr\{\xi(\omega) \geqslant x\}\, dx - \int_{0}^{\infty} Cr\{\xi(\omega) \leqslant x\}\, dx \right] P(d\omega). \tag{5.3}$$

Definition 5.2 Let ξ be a normal random fuzzy variable defined on a possibility space $(\Theta, \mathcal{P}(\Theta), Pos)$ and assume that $E(\xi) < \infty$. Then, the variance $Var(\xi)$ is defined as the expected value of a random fuzzy variable $(\xi - E(\xi))^2$, i.e.,

$$Var(\xi) = E\left((\xi - E(\xi))^2\right). \tag{5.4}$$

Based on a fuzzy random variable, upper and lower distribution functions may be estimated. These functions characterise uncertainty of the analysed variable. It is known, that each fuzzy variable $\tilde{Z}$ with possibility distribution π induces a random set [32]. Let α-levels of the said variable be marked with π^{α}. Focal elements of a random set generated by a fuzzy variable $\tilde{Z}$ are α-levels $(\pi^{\alpha_j})_{j=1,2,\ldots,q}$, where $\alpha_0 = \alpha_1 = 1 > \alpha_2 > \cdots \alpha_q > \alpha_{q+1} = 0$. Values $(v_j = \alpha_j - \alpha_{j+1})_{j=1,\ldots,q}$ constitute probability mass of the generated random set. Every random set induces *Plausibility* (*Pl*) and *Belief* functions (*Bel*). Based on the defined functions *Pl* and *Bel*, one may determine the upper $\overline{F}(x)$ and lower $\underline{F}(x)$ distribution function in compliance with formulae (3.8) and (3.9).

5.3 Computing an Operation Profit

In order to illustrate the effectiveness of the proposed methods, below are presented the results of calculations made for a simple model problem. The methods were verified on the example of calculation of operating profit for a metallurgical industry enterprise. The calculation was performed for the production system presented in Fig. 5.1.

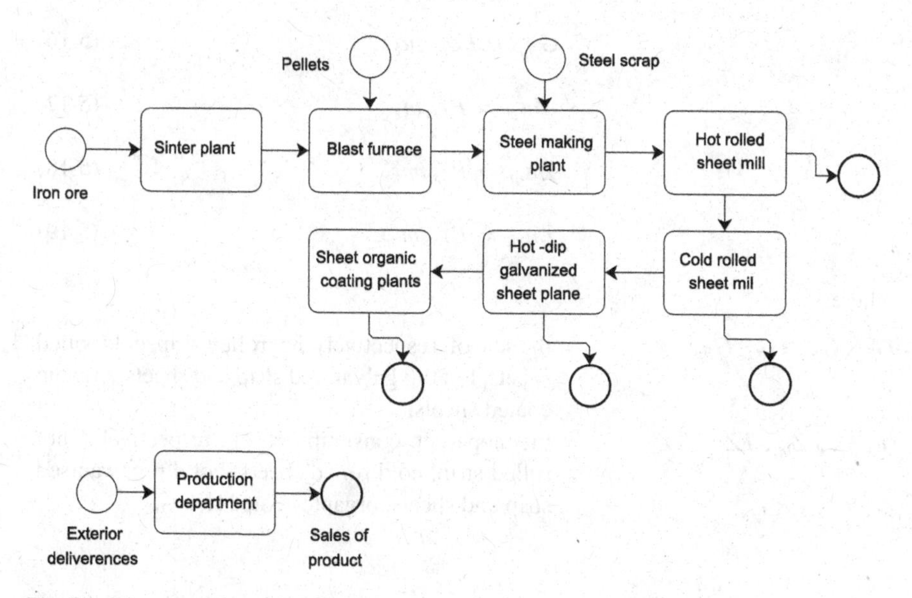

Fig. 5.1 Diagram of the analysed production system

Operating profit can be expressed by the formulae below.

$$\begin{aligned} ZO = {} & c_{bg}G_{bg} + c_{bz}G_{bz} + c_{bo}G_{bo} + c_{bp}G_{bp} \\ & - k_{sp}Pr_{sp} - k_{su}Pr_{su} - k_{st}Pr_{st} \\ & - k_{bg}Pr_{bg} - k_{bz}Pr_{bz} - k_{bo}Pr_{bo} - k_{bp}Pr_{bp} \\ & - Zu_{zl}c_{zl} - Zu_{gr}c_{gr} - Zu_{ru}c_{ru} - kf, \end{aligned} \tag{5.5}$$

$$Pr_{sp} = Pr_{su}m_{sp}, \tag{5.6}$$

$$Pr_{su} = Pr_{st}m_{su}, \tag{5.7}$$

$$Pr_{st} = Pr_{bg}m_{pl}, \tag{5.8}$$

$$Pr_{bg} = Pr_{bz}m_{bg} + G_{bg}, \tag{5.9}$$

$$Pr_{bz} = Pr_{bo}m_{bz} + G_{bz}, \tag{5.10}$$

$$Pr_{bo} = Pr_{bp}m_{bo} + G_{bo}, \tag{5.11}$$

$$Pr_{bp} = G_{bp}, \tag{5.12}$$

$$G_{bg} = JZ_{bg}u_{bg}, \tag{5.13}$$

$$G_{bz} = JZ_{bz}zu_{bz}, \tag{5.14}$$

$$G_{bo} = JZ_{bo}zu_{bo}, \tag{5.15}$$

$$G_{bp} = JZ_{bp}zu_{bp}, \tag{5.16}$$

$$Zu_{zl} = Pr_{st}m_{zl}, \tag{5.17}$$

$$Zu_{gr} = Pr_{su}m_{gr}, \tag{5.18}$$

$$Zu_{ru} = Pr_{sp}m_{ru}, \tag{5.19}$$

where:

$G_{bg}, G_{bz}, G_{bo}, G_{bp}$ the sale of, respectively, hot rolled strip, cold rolled sheets, hot dip galvanised strip and sheets, organic coated sheets,

$JZ_{bg}, JZ_{bz}, JZ_{bo}, JZ_{bp}$ the apparent consumption of, respectively, hot rolled strip, cold rolled sheets, hot dip galvanised strip and sheets, organic coated sheets,

$u_{bg}, u_{bz}, u_{bo}, u_{bp}$	the market share of, respectively, hot rolled strip, cold rolled sheets, hot dip galvanised strip and sheets, organic coated sheets,
$c_{ru}, c_{gt}, c_{zl}, c_{bg}, c_{bz}, c_{bo}, c_{bp}$	the price of, respectively, iron ore, pellets, steel scrap, hot rolled strip, cold rolled sheets, hot dip galvanised strip and sheets, organic coated sheets,
$m_{sp}, m_{su}, m_{pl}, m_{bg}, m_{bz}, m_{bo}$	respectively, the sinter consumption ratio per tonne of pig-iron, pig-iron consumption ratio per tonne of continuous casting stands, continuous casting stands consumption ratio per tonne of hot rolled sheets, hot rolled sheets consumption ratio per tonne of cold rolled sheets, cold rolled sheets consumption ratio per tonne of galvanised sheets, galvanised sheets consumption ratio of per tonne of organic coated sheets,
m_{zl}, m_{gr}, m_{ru}	the ratio of, respectively, scrap consumption per tonne of steel, ratio of pellets consumption per tonne of pig-iron, iron ore consumption ratio per tonne of sinter,
$Pr_{sp}, Pr_{su}, Pr_{st}, Pr_{bg}, Pr_{bz}, Pr_{bo}, Pr_{bp}$	the production of, respectively, sinter, pig iron, continuous casting stands, hot rolled strip, cold rolled sheets, hot dip galvanised strip and sheets, organic coated sheets,
$k_{sp}, k_{su}, k_{st}, k_{bg}, k_{bz}, k_{bo}, k_{bp}$	the properly adjusted unit variable cost[1] of, respectively, sinter, pig iron, continuous casting stands, hot rolled strip, cold rolled sheets, hot dip galvanised strip and sheets, organic coated sheets,
kf	company's fixed costs,
$Zu_{zl}, Zu_{gr}, Zu_{ru}$	scrap, pellets and iron ore consumption, respectively.

Figure 5.2 shows prices of metallurgical products manufactured by analysed company and prices of iron ore, pellets and steel scrap in 1992–2011. Apparent consumption of metallurgical products manufactured by analysed company is depicted in Fig. 5.3.

Trapezoidal fuzzy numbers specifying forecast of parameters for calculating the operating profit for the analysed company are presented in Table 5.1. The relations between prices of the analysed ranges of steel products, prices of iron ore, pellets and scrap as well as the apparent consumption of particular product ranges were expressed by means of the interval regression model. The coefficients of the regression equations were estimated using the method described in Sect. 1.3.2. Table 5.2 presents the

[1]This cost does not account for the values of used steel products manufactured in previous stages of the cycle as well as value of used raw materials, corrections are done in order to avoid multiple calculation of the same cost components during calculation of profit, according to formula (4.1).

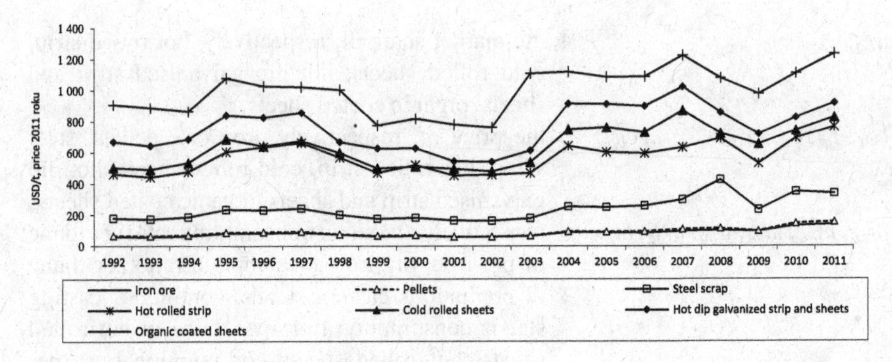

Fig. 5.2 Prices of metallurgical products manufactured by analysed company and prices of iron ore, pellets and steel scrap in 1992–2011

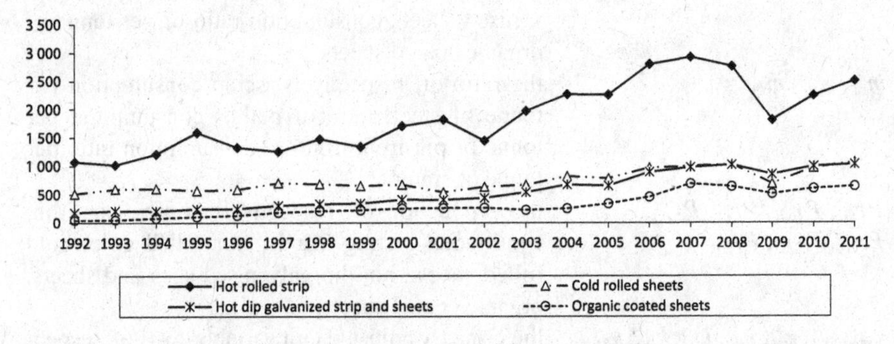

Fig. 5.3 Apparent consumption of metallurgical products manufactured by analysed company

exemplary coefficients of the interval regression equations characterising relations between prices of products manufactured by analysed producer and prices of iron ore, pellets and scrap.

The value of the fixed cost was adopted at the level of USD 315 090 thou/year. Adjusted unit variable processing cost for particular product ranges were also adopted at the levels given below:

Product	Sinter	Pig iron	Continuous casting stands	Hot rolled strip	Cold rolled sheets	Hot dip galvanised strip and sheets	Organic coated sheets
Adjusted unit variable processing cost, USD/t	16.6	153.8	25.4	28.4	28	116.7	175.3

The following market share values of particular product ranges were adopted in the calculations:

Hot rolled strip	Cold rolled sheets	Hot dip galvanised strip and sheets	Organic coated sheets
42.50 %	40.00 %	46.00 %	45.00 %

Table 5.1 Trapezoidal fuzzy numbers representing forecasts of products and raw material prices, material consumption indicators and apparent consumption of metallurgical products

Price	Trapezoidal fuzzy numbers, USD/tonne
Iron ore	(111.7, 120.0, 133.3, 141.7)
Pellets	(125.0, 133.3, 146.7, 156.7)
Steel strap	(313.3, 320.0, 336.7, 345.0)
Hot rolled strip	(666.7, 680.0, 711.7, 728.3)
Cold rolled sheets	(715.0, 730.0, 763.3, 781.7)
Hot dip galvanised strip and sheets	(805.0, 821.7, 860.0, 880.0)
Organic coated sheets	(1 080.0, 1 101.7, 1 153.3, 1 175.0)
Material consumption indicators	Trapezoidal fuzzy numbers, tonne/tonne
Iron ore—sinter	(0.918, 0.920, 0.920, 0, 922)
Sinter—pig iron	(1.352, 1.354, 1.359, 1.362)
Pellets—pig iron	(0.338, 0.339, 0.340, 0.341)
Scrap—continuous casting stands	(0.269, 0.276, 0.279, 0.288)
Pig iron—continuous casting stands	(0.855, 0.860, 0.870, 0.875)
Continuous casting stands—hot rolled strip	(1.058, 1.064, 1.075, 1.078)
Hot rolled strip—cold rolled sheets	(1.105, 1.111, 1.124, 1.130)
Cold rolled sheets—hot dip galvanised strip and sheets	(1.010, 1.020, 1.026, 1.031)
Hot dip galvanised strip and sheets—organic coated sheets	(0.998, 0.999, 1.000, 1.001)
Apparent consumption	Trapezoidal fuzzy numbers, thousands of tonnes
Hot rolled strip	(2 405.4, 2 704.0, 2 704.0, 3 033.6)
Cold rolled sheets	(1 018.2, 1 162.3, 1 162.3, 1 293.0)
Hot dip galvanised strip and sheets	(1 008.4, 1 147.9, 1 147.9, 1 287.4)
Organic coated sheets	(625.9, 708.4, 708.4, 791.2)

The operating profit was calculated using different methods of mathematical operations. The obtained results are presented in Figs. 5.4, 5.5, 5.6, 5.7, 5.8 and 5.9. Profits labelled with numbers I–IV have been calculated using various methods of implementation of arithmetic operations on fuzzy numbers. Operating profits V and VI were calculated using the hybrid propagation method. At the calculation of this profits the apparent steel consumption of individual metallurgic product mixes one expressed in the form of probability distributions. Table 5.3 presents probability density function parameters presenting apparent steel consumption for products produced by the company. Remaining parameters burdened an uncertainty in case of these calculations were expressed in the form of numbers fuzzy presented in Table 5.1. Table 5.4

Table 5.2 Interval regression equations coefficients depicting interrelations between prices of particular product ranges produced by the manufacturer in question prices of raw materials

Independent variable		Dependent variable						
		Iron ore	Pellets	Steel scrap	Hot rolled strip	Cold rolled sheets	Hot dip galvanised strip and sheets	Organic coated sheets
Iron ore	a_1		[0.80, 0.89]	[0.20, 0.34]	[0.17, 0.26]	[0.10, 0.20]	[0.06, 0.17]	[0.07, 0.16]
	a_2		[6.87, 7.69]	[13.00, 22.39]	[−46.54, −30.09]	[−18.17, −9.18]	[−7.70, −2.80]	[−48.12, −21.79]
Pellets	a_1	[1.11, 1.25]		[0.24, 0.42]	[0.19, 0.30]	[0.11, 0.24]	[0.07, 0.21]	[0.08, 0.19]
	a_2	[−8.30, −7.37]		[8.46, 14.91]	[−63.98, −39.89]	[−32.08, −15.30]	[−23.86, −7.89]	[−67.82, −28.57]
Steel strap	a_1	[1.46, 4.07]	[1.32, 3.42]		[0.40, 0.87]	[0.21, 0.71]	[0.32, 0.60]	[0.13, 0.56]
	a_2	[5.16, 14.37]	[14.65, 38.10]		[−172.16, −80.04]	[−86.59, −26.03]	[−73.93, −38.69]	[−178.41, 42.87]
Hot rolled strip	a_1	[3.48, 4.51]	[2.90, 3.80]	[1.01, 1.36]		[0.60, 0.79]	[0.48, 0.66]	[0.39, 0.66]
	a_2	[211.97, 275.29]	[237.92, 311.40]	[247.72, 334.56]		[112.16, 148.39]	[112.61, 155.43]	[31.68, 53.32]
Cold rolled sheets	a_1	[3.55, 5.80]	[3.04, 4.83]	[1.12, 1.70]	[0.87, 1.41]		[0.75, 0.92]	[0.57, 0.96]
	a_2	[191.11, 312.26]	[221.35, 351.44]	[239.84, 364.71]	[−17.12, −10.59]		[−2.65, −2.18]	[−159.73, −95.30]
Hot dip galvanised strip and sheets	a_1	[3.59, 5.76]	[3.14, 4.89]	[1.19, 1.78]	[0.96, 1.50]	[0.98, 1.22]		[0.60, 1.07]
	a_2	[292.01, 468.72]	[318.79, 497.23]	[331.01, 492.93]	[45.67, 71.29]	[57.88, 72.08]		[−87.59, −48.89]
Organic coated sheets	a_1	[3.96, 6.14]	[3.28, 5.20]	[1.16, 1.79]	[0.35, 0.87]	[0.81, 1.31]	[0.63, 1.12]	
	a_2	[458.75, 712.03]	[482.39, 765.86]	[511.55, 790.07]	[385.57, 941.97]	[248.18, 403.48]	[236.97, 420.89]	

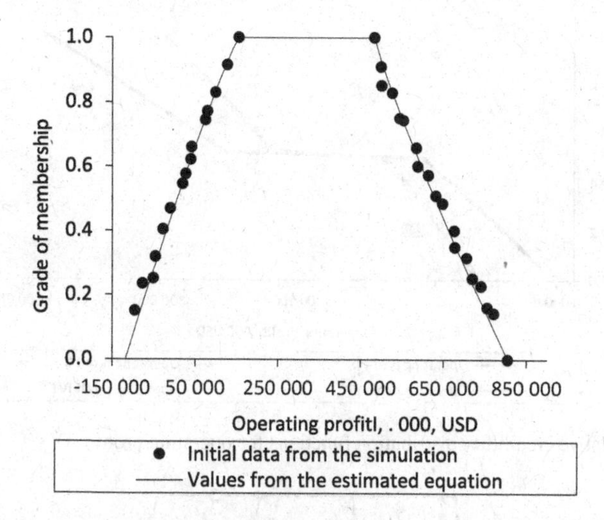

Fig. 5.4 Fuzzy numbers representing the operating profit I

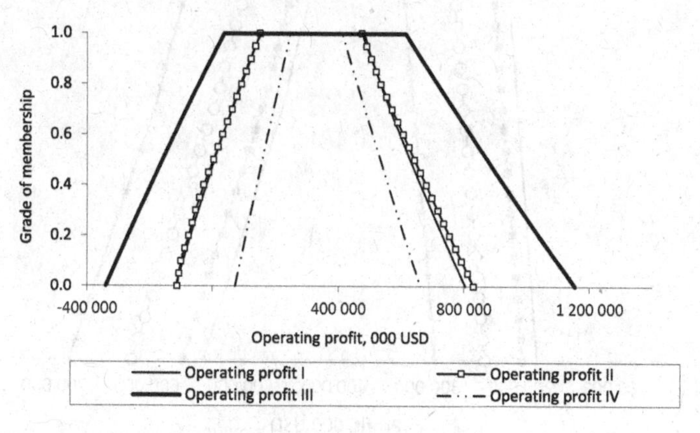

Fig. 5.5 Fuzzy numbers representing the operating profit I–IV

presents a matrix of correlation of apparent consumption of particular product ranges manufactured by a producer.

The value of all coefficients of correlation was significant at testing by means of the t-Student statistics for the significance level $\alpha = 0.05$.

The operating profit I was calculated using simulation of fuzzy systems (method described in Sect. 1.3.1) for execution of arithmetic operations on interactive fuzzy numbers. The operating profit II one calculated using non-linear programming (method described in Sect. 1.4.1) for execution of arithmetic operations on interactive fuzzy numbers. Additionally, for comparison one calculated the operating profit III and IV. The operating profit III one calculated on the assumption that did not exist dependencies among prices of individual product mixes and prices of raw materials and dependencies among the apparent consumption of individual metallur-

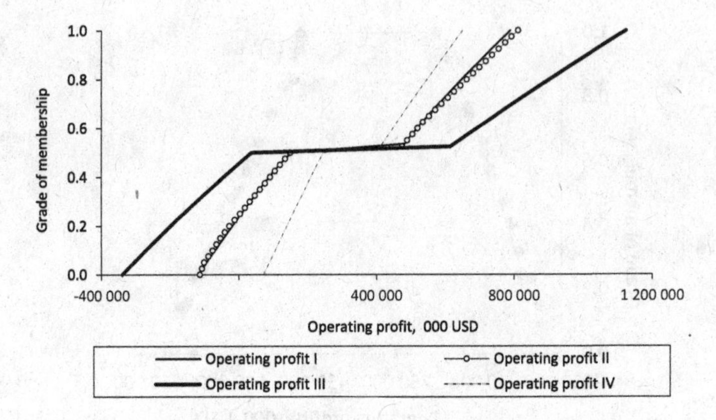

Fig. 5.6 Cumulative credibility distribution functions for operating profit I–IV

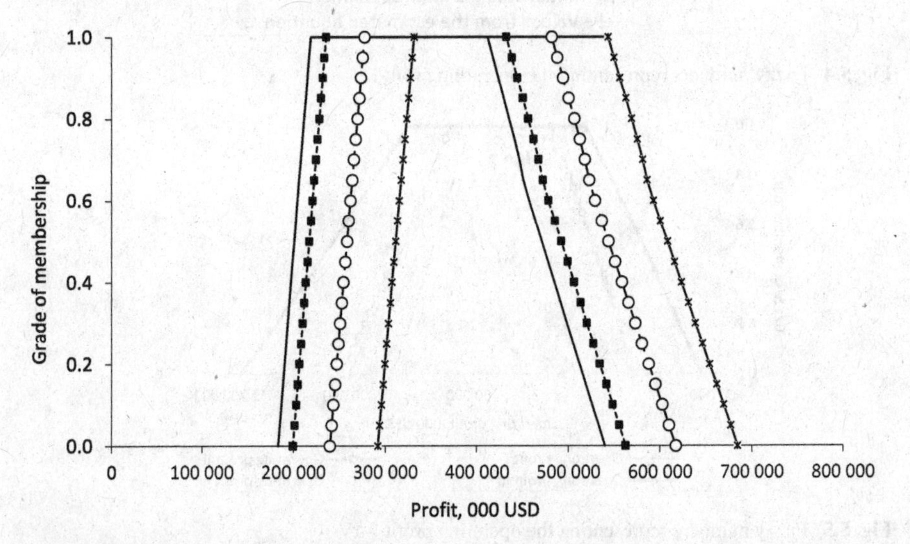

Fig. 5.7 Example fuzzy numbers depicting operating profit calculated in the selected iterations of computer simulation

gic product mixes. The operating profit III was calculated as follows: the production cost of each product, from organic coated sheets to pig iron, was calculated using the Eqs. (1.29) and (1.30), the revenue was calculated using Eqs. (1.29) and (1.30), and the operating profit was calculated by subtracting the production costs from the revenue, according to the Eq. (1.38). At this stage one applied the constrained subtraction. The use of the Eq. (1.38) would mark, the subtraction of costs calculated for the greatest possible production from the income calculated for least sales and vice versa.

The operating profit IV one calculated on the assumption that existed strong, functional dependencies among prices of individual product mixes and prices of raw

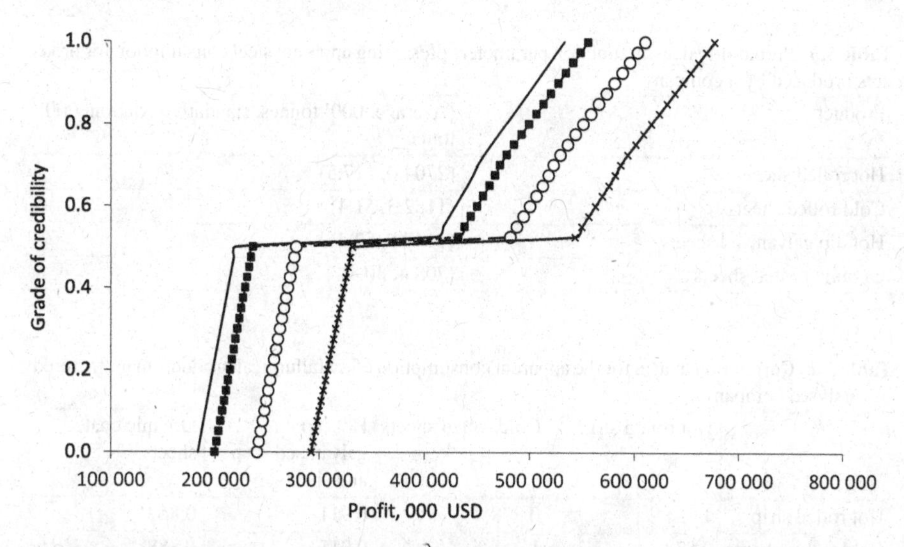

Fig. 5.8 Credibility distribution functions depicting operating profit calculated in selected iterations of computer simulation

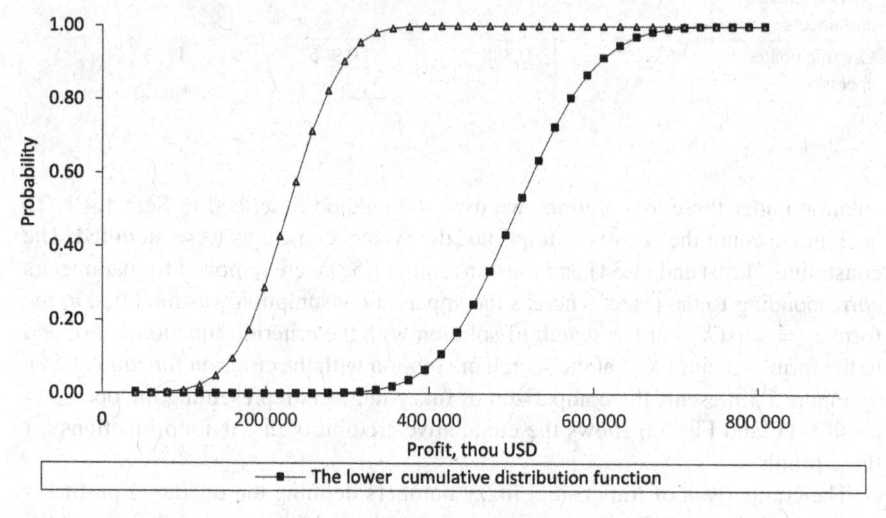

Fig. 5.9 Upper and lower distribution functions depicting operating profit

materials and dependencies among the apparent consumption of individual metallurgic product mixes. The operating profit IV one calculated on the assumption that the highest possible prices of the various product ranges correspond to the highest prices of metallurgical raw materials and the lowest possible prices of individual assortments of steel products correspond to the lowest prices of raw materials. Similarly one assumed that the apparent consumption of all metallurgic product mixes accepts highest possible values or lowest possible values. To implement the cal-

Table 5.3 Probability density function parameters presenting apparent steel consumption for products produced by a company

Product	(Average, 000' tonnes, standard deviation, 000' tonnes)
Hot rolled sheets	(2704.0, 117.5)
Cold rolled sheets	(1162.3, 51.4)
Hot dip galvanised sheets	(1147.9, 52.4)
Organic coated sheets	(708.4, 30.8)

Table 5.4 Correlation matrix for the apparent consumption of metallurgical products manufactured by analysed company

	Hot rolled strip	Cold rolled sheets	Hot dip galvanised strip and sheets	Organic coated sheets
Hot rolled strip	1	0.878	0.911	0.863
Cold rolled sheets	0.878	1	0.915	0.888
Hot dip galvanised strip and sheets	0.911	0.915	1	0.966
Organic coated sheets	0.863	0.888	0.966	1

culation under these assumptions we used the method described in Sect. 1.4.1. To take into account the above assumptions, the system constrains were modified. The constraints (1.53) and (1.54) and the constraint (1.52) were removed for parameters corresponding to the price, whereas the apparent consumption was modified to the form $x_i = \sup(X_i^\alpha)$ at the search of solution with the criterion function (1.50) and to the form $x_i = \inf(X_i^\alpha)$ at the search of solution with the criterion function (1.51).

Figure 5.5 presents the comparison of fuzzy numbers representing the operating profit I–IV and Fig. 5.6 shows the cumulative credibility distribution functions for these profits.

The comparison of trapezoidal fuzzy numbers defining the operating profit I is presented in Fig. 5.4. The figure presents data obtained from the simulation and the data determined thanks to the appraisal value of parameters of the equation:

$$\mu(x) = \begin{cases} 1 - \frac{b-x}{b-a} & \text{for } a \le x \le b \\ 1 & \text{for } b \le x \le c \\ 1 - \frac{x-c}{c-b} & \text{for } c \le x \le d \\ 0 & \text{otherwise} \end{cases},$$

where a, b, c, d—parameters of the equation.

Figures 5.7 and 5.8 depict example results of calculations of the operating profit with the use of the hybrid method (operating profit V). Figure 5.7 presents four fuzzy numbers characterising operating profit calculated in the selected iterations of computer simulation and Fig. 5.8 presents respective credibility distribution functions. The average value of operating profit was *USD* 361333.6 thou. and standard semi-deviation *USD* 94303.6 thou. Figure 5.9 depicts pessimistic and optimistic cumulative distribution functions resulting from the said computations.

Additionally, calculations were effected without consideration of the correlation of economic parameters (operating profit VI). In this case, the average of operating profit obtained was 315488.9 and standard semi-deviation was 308475.5. Figure 5.10 presents the resulting optimistic and pessimistic cumulative distribution functions.

The average and standard semi-deviation of the operating profit for different variants of calculations one presented in Table 5.5.

Fuzzy numbers and cumulative credibility distribution functions presented in Figs. 5.4, 5.5 and 5.6 show that the application of different variants of arithmetic operations give different results. The assumption adopted for the calculation of the operating profit III and IV results in too narrow or too broad compartments for α-levels of fuzzy numbers representing the result of arithmetic operations. They are unreal and impossible to obtain in reality. This results in a suitably large or small values of the standard semi-deviation of operating profit. The application of the method for performing arithmetic operations on fuzzy numbers that was described in Sects. 1.3.1 and 1.4.1 eliminates the deficiencies of arithmetic operations performed according to the Eqs. (1.29), (1.31) and (1.38). Thanks to this methods, it is possible to take into account all interactivities between the analysed quantities. However, in the case of typical links appearing between economical parameters in

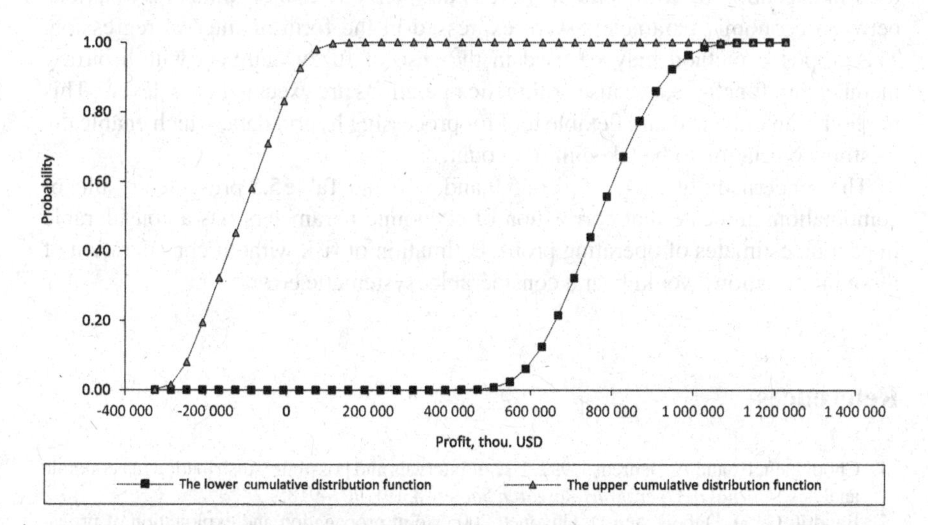

Fig. 5.10 Upper and lower distribution functions depicting operating profit obtained in the variant of calculations not taking into account correlation of parameters of the economic calculation

Table 5.5 The average and standard semi-deviation of the operating profit for different variants of calculations

	Average, 000' USD	Std. semi-dev., 000' USD
Operating profit I	307 918.00	217 917.70
Operating profit II	305 070.40	214 469.40
Operating profit III	330 511.30	348 214.80
Operating profit IV	309 084.40	123 603.30
Operating profit V	331333.6	94 303.60
Operating profit VI	315488.9	308475.5

a company, calculating the values of arithmetic expressions characterising financial indices requires the application of nonlinear programming methods, indeed, this complicates the process of calculation. Moreover, in case of application of the methods of non-linear programming (the method described in Sect. 1.3.1), estimation of a large number of equations of fuzzy linear regression is necessary. With a larger number of interactive parameters, it may complicate the calculation process. This defect is not present with the method using simulations of fuzzy systems. However, in case of a large number of analysed parameters, the correspondingly large number of replications is required, which extends the calculation process.

Proposed hybrid method facilitates processing of hybrid data, taking into account the correlation between economic parameters. The procedure for processing such data combines stochastic simulation with Zadeh's extension principle (the method for execution of arithmetic operations on fuzzy numbers). Non-linear programming was used in execution of arithmetic operations on interactive fuzzy numbers. Relations between economic parameters were expressed in the form of interval regression. The proposed method may be used in the case of fuzzy numbers with arbitrary membership functions, because arithmetic operations are executed on α-levels. This method is an universal and flexible tool for processing hybrid data, which enable the existing correlation to be taken into account.

The concerned Figs. 5.4, 5.5, 5.6, 5.9 and 5.10 and Table 5.5 presented results of computations indicate that correlation of economic parameters has a considerable impact on estimates of operating profit. Estimation of risk without consideration of these interrelations would bear a considerable, systematic error.

References

1. Choobineh, F., and A. Behrens. 1992. Use of intervals and possibility distribution in economic analysis. *Journal of Operations Research Society* 43(9): 907–918.
2. Baudrit, C., D. Dubois, and D. Guyonet. 2006. Joint propagation and exploitation of probabilistic and possibilistic information in risk assessment. *IEEE Transaction on Fuzzy Systems* 14(5): 593–607.

3. Gupta, C.P. 1993. A note on the transformation of possibilistic information into probabilistic information for investment decisions. *Fuzzy Sets and Systems* 56(2): 175–182.
4. Kuchta, D. 2001. *Miękka matematyka w zarządzaniu: zastosowanie liczb przedziałowych i rozmytych w rachunkowości zarządczej*. Wrocław: Oficyna Wydawnicza Politechniki Wrocławskiej.
5. Pluta, W., and T. Jajuga. 1995. Inwestycje: Capital Budgeting—Budżetowanie Kapitałowe. Fundacja Rozwoju Rachunkowości w Polsce.
6. Dittmann, P. 2008. *Prognozowanie w przedsiębiorstwie: metody i ich zastosowanie*. Wolters Kluwer.
7. Kuchta, D. 2000. Fuzzy capital budgeting. *Fuzzy Sets and Systems* 111(3): 367–385.
8. Mohamed, S., and A.K. McCowan. 2001. Modelling project investment decisions under uncertainty using possibility theory. *International Journal of Project Management* 19(4): 231–241.
9. Ferson, S. 1996. What Monte Carlo method cannot do. *Human and Ecological Risk Assessment: An International Journal* 2(4): 990–1007.
10. Georgescu, I., and J. Kinnunen. 2011. Modeling the risk by credibility theory. In *Proceedings of the 3rd international IEEE conference on advanced management science (ICAMS 2011), Kuala Lumpur, Malaysia, International proceedings of economics development and research*, vol. 19, 15–19.
11. Chiu, C.-Y., and S.C. Park. 1994. Fuzzy cash flow analysis using present worth criterion. *The Engineering Economist* 39(2): 113–138.
12. Rębiasz, B. 2013. Selection of efficient portfolios-probabilistic and fuzzy approach, comparative study. *Computers and Industrial Engineering* 64(4): 1019–1032.
13. Rębiasz, B. 2007. Fuzziness and randomness in investment project risk appraisal. *Computers and Operations Research* 34(1): 199–210.
14. Ferson, S., and L.R. Ginzburg. 1996. Difference method are needed to propagate ignorance and variability. *Reliability Engineering and System Safety* 54(2–3): 133–144.
15. Cooper, J.A., S. Ferson, and L. Ginzburg. 1996. Hybrid processing of stochastic and subjective uncertainty data. *Risk Analysis* 16(6): 785–791.
16. Ward, T.L. 1985. Discounted fuzzy cash flow analysis. In *1985 Fall Industrial engineering* 476–481.
17. Buckley, J.J. 1992. Solving fuzzy equations in economics and finance. *Fuzzy Sets and Systems* 48(3): 289–296.
18. Calzi, M.L. 1990. Toward a general setting for the fuzzy mathematics of finance. *Fuzzy Sets and Systems* 35(3): 265–280.
19. Esogbue, A.O., and W.E. Hearnes II. 1998. On replacement models via a fuzzy set theoretic framework. *IEEE Transactions on Systems Manufacturing and Cybernetics—Part C: Applications and Reviews* 28(4): 549–560.
20. Kahraman, C., D. Ruan, and E. Tolga. 2002. Capital budgeting techniques using discounted fuzzy versus probabilistic cash flows. *Information Science* 142(1–4): 57–76.
21. Rębiasz, B., B. Gaweł, and I. Skalna. 2014. Capital budgeting of interdependent projects with fuzziness and randomness. In *Information systems architecture and technology*, ed. Zofia Wilimowska et al, 125–135.
22. Rębiasz, B., B. Gaweł, and I. Skalna. 2015. Fuzzy multi-attribute evaluation of investments. In *Advances in ICT for business, industry and public sector: ABICT'13 (4th international workshop on Advances in business ICT)*, eds. Tomasz Pełech-Pilichowski Maria Mach-Król, and Celina M. Olszak, 141–156.
23. Rębiasz, B., B. Gawel, and I. Skalna. 2014. Hybrid framework for investment project portfolio selection. In *Federated conference on computer science and information systems* 1123–1228.
24. Smets, P. 1990. Constructing the pignistic probability function in a context of uncertainty. In *Proceedings of the fifth annual conference on uncertainty in artificial intelligence*, ed. Henrion M., E.D. Schachter, L.N. Kanal, and J.F. Lemmer. UAI'89, 29–39. Amsterdam: North-Holland Publishing Co.
25. Shafer, G. 1976. *A mathematical theory of evidence*. Princeton: Princeton University Press.

26. Guyonnet, D., B. Bourgine, D. Dubois, H. Fargier, B. Côme, and P.J. Chilès. 2003. Hybrid approach for addressing uncertainty in risk assessment. *Journal of Environmental Engineering* 129(1): 68–76.
27. Kaufmann, A., and M.M. Gupta. 1985. *Introduction to fuzzy arithmetic: Theory and application.*, Electrical-computer science and engineering series New York: van Nostrand Reinhold Company.
28. Puri, M.L., and D.A. Ralescu. 1985. The concept of normality for fuzzy random variables. *Annals of Probability* 13(4): 1371–1379.
29. Liu, Y.-K., and B. Liu. 2003. Fuzzy random variables: A scalar expected value operator. *Fuzzy Optimization and Decision Making* 2(2): 143–160.
30. Liu, Y.-K., and B. Liu. 2003. Expected value operator of random fuzzy variable and random fuzzy expected value models. *International Journal of Uncertainty, Fuzziness and Knowledge-Based Systems* 11(2): 195–215.
31. Liu, Y.-K., and B. Liu. 2002. Expected value of fuzzy variable and fuzzy expected value models. *IEEE Transactions on Fuzzy Systems* 10(4): 445–450.
32. Zadeh, L.A. 1978. Fuzzy sets as a basis for a theory of possibility. *Fuzzy Sets and Systems* 1(1): 3–28.

Chapter 6
Application of Fuzzy Theory in Steel Production Planning and Scheduling

Abstract This chapter describes the application of fuzzy sets to planning and scheduling of production in the steel industry. Primarily, the problem of steel grade assignment to customer' orders is analysed. Fuzzy sets are used to reduce the variety of potential steel grades and to describe characteristic of materials by decision makers. Next, fuzzy logic systems for steel production scheduling are examined. Fuzzy parameters and fuzzy constraints are used to describe some aspects of the steel production process, with a special respect to the continuous casting. Finally, the cooperation of steel production planning between different shops using a multi-agent approach and fuzzy sets is discussed and the practical example of a genetic algorithm applied to solve a fuzzy lot-sizing problem for a continuous casting planning agent is presented.

Typical steelworks consists of a steel melting shop (furnace), at least one continuous casting line, a hot strip mill and, optionally, further processing facilities, such as cold rolling mill or pickling line. The steel is molten in furnaces and transferred directly to continuous casters. Caster shapes molten steel into slabs with different widths and chemical composition (grades). Finally, the slabs are rolled in the hot strip mill to produce steel coils.

Planning and scheduling of steel production is a very demanding challenge due to the complex nature of a steel making process. Many parameters of this process are expressed by normative values that may differ from the real values occurring in a given production circumstances. Decision makers must take into account some level of uncertainty of parameters and adjust the process according to their own knowledge and experience. The most important aspects of the steel production process that must be reflected in the production schedules are the following [1]:

- At the steel-making stage: orders change, machines failures, smelting time exceeding the prescribed limit etc.
- At the continuous caster stage: information concerning the true weight of the ladle contents, heats (a fixed tonnage of molten steel) arrive randomly and dynamically, heats of steel having the wrong chemistry, steel leak appears in the continuous caster, machine failure etc.

© Springer International Publishing Switzerland 2015
I. Skalna et al., *Advances in Fuzzy Decision Making*,
Studies in Fuzziness and Soft Computing 333,
DOI 10.1007/978-3-319-26494-3_6

- At the hot strip mill stage: slab quality is not up to standard, slabs are delayed, slabs are backed up in the hot strip mill, a new high-priority order is introduced, an order is cancelled, machine failure etc.

The uncertainty concerns not only the production process itself, but also the customer demand for specific types and grades of steel products, quality issues, and logistics (procurement, transport between successive stages, delivery to the final customer).

Zarandi and Ahmadpour [2] enumerate general characteristic of the steel industry in the context of planning and scheduling tasks:

- Steel production is a multi-stage process, logically and geographically distributed, involving a variety of production processes
- There are different problems and different problem solving methods for different steel making stages.
- Manual techniques used in steel making processes are based on the know-how and the professional experience of expert people who have worked in the plant for years.
- The output of some stages is the input for some other stages, so the integrated process is necessary.
- Making contracts requires negotiation between buyer and supplier.

All the above factors cause that steelworks usually do not use a fully computerised planning and scheduling system, especially at the shop floor level. Nevertheless, some individual solutions supporting planning and scheduling of the production process at various stages of decision making has been successfully developed in many steelworks. Tang et al. [1] have classified the methods used in such solutions into the following groups:

- Operations research methods in which linear programming or mixed integer programming model is built and then some exact or heuristic method is used in ordered to solve it.
- Artificial intelligence methods including expert systems, computational intelligence and constraint satisfaction approach.
- Visualisation techniques in human-machine coordination systems.
- Multi-agent approach.

In both artificial intelligence and multi-agent approaches fuzzy sets or fuzzy logic is often applied in order to describe uncertain parameters of a steel production process. The following sections present the application of fuzzy theory and fuzzy logic to the planning and scheduling process, beginning with the problem of grade assignment, job scheduling taking into account fuzzy due dates, and coordination of the plans by exchanging messages between agents. Finally a detailed optimisation model with fuzzy parameters for the continuous caster agent has been shown and a genetic algorithm able to solve it in an efficient way has been proposed.

6.1 Fuzzy Grade Assignment

Customer orders for steel products are characterised by two major parameters: dimension and steel grade. Steel grades are usually standardised according to either intended use and mechanical properties of the steel or its chemical composition.

In such standards many of the mechanical parameters and chemical characteristics are given by their lower and upper bounds (or one of them). For example the steel graded identified as SAE 304 gives the following ranges of the chemical components: Cr 23–26%, Ni 19–22 %, C 0.25 %, Mn 2 %, Si 1.5–3 %, P 0.045 % and S 0.03 %.

Due to inevitable uncertainties in the steel production process (especially at the steelmaking level), it is hardly possible to produce steel with a precisely predetermined quality and to maintain the same quality for all orders placed by a customer. Therefore the quality of the steel assigned to the product is usually better than requested by the customer. This is done not only to protect the quality requirements of the customer, but also to improve the efficiency of the steel making process, e.g., to consume all the steel that was melted in the furnace, or to achieve a maximal throughput of the continuous casters.

The uncertainty concerning chemical composition results from the fact, that the steelmaking process for different grades is performed on the same units and in particular from [3]:

- residuals in an electric arc furnace remain in the wall and will be assimilated by the subsequent heat,
- two heats with different steel grades casted one after another are mixed in the tundish.

To avoid the above problems Vasko et al. [4] and later Woodyatt et al. [5] proposed a method based on the concept of fuzzy sets that reduced the variety of potential steel grades. The idea was that instead of a single grade specified in a customer order, it is assigned a set of possible grades that would meet customer's requirements. Figure 6.1 illustrates that requirements for carbon and manganese can be satisfied by five different steel grades.

The method works as follows. First, a set of potential grades that can be used to satisfy customer's requirements is assigned to the order and then the grade that will be sufficient to produce all the orders is selected. The probability that the grade will meet the specification of the customer's order is described by fuzzy sets. The selection of one grade for a collection of customer orders is performed by a set covering approach.

The complete collection of N grades is identified as:

$$G = \{G_i\}, \forall j = 1, 2, \ldots, N \tag{6.1}$$

A collection of M customer orders to which a steel grade must be assigned:

$$S = \{S_i\}, \forall i = 1, 2, \ldots, M \tag{6.2}$$

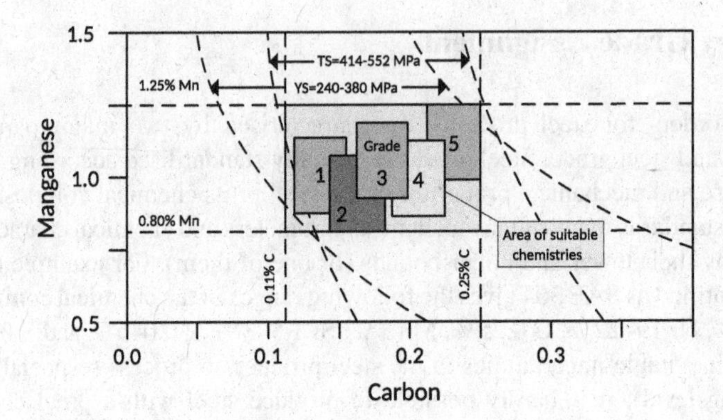

Fig. 6.1 Grade assignment on the basis of fuzzy carbon and manganese requirements [4]

Finally, a collection of fuzzy subsets of G, such as $A(i)$ measures the ability of a grade in the list G to meet the specification of customers order S_i.

$$A = \{A(i)\}, \forall i = 1, 2, \ldots, M \tag{6.3}$$

The membership function f for each $A(i)$depends on three components:

- the ability of a grade i in the list G to meet all mechanical property requirements of customer order S_i
- the ability of a grade i in the list G to meet all chemical requirements of customer order S_i
- desirability of producing a grade i in the list G relative to other grades in the list.

and is defined as follows:

$$\begin{aligned} f_{A(i,K)}(G_i) &= \text{Prob}[m(i, K) \leq X(j, K) \leq M(i, K)] \\ \forall j &= 1, 2, \ldots, N, \forall i = 1, 2, \ldots, M, \forall K = 1, 2, \ldots, U \end{aligned} \tag{6.4}$$

where U is the number of mechanical properties, $m(i, K)$ and $M(i, K)$ are, respectively, the minimum and maximum level of mechanical property K for order S_i.

The solution proposed by Vasko et al. [4] has been implemented in Bethlehem Steel (USA) as the MGAP module (metallurgical grade assignment program) within the production planning and control (PPC) system. Limitation of steel grades produced by the steelworks contributed to better utilisation of continuous casting lines, but also to reduction of molten steel necessary to satisfy customers' orders.

In [6] Wang and Chang used fuzzy sets to solve a similar problem of assignment tool steel to customer's order. This time, however, the fuzzy sets were used to express the importance of criteria for decision makers. The following characteristic of materials were considered:

- non-deforming properties,
- safety in hardening,
- toughness,
- resistance to softening effect of heat,
- wear resistance,
- machinability.

All those criteria are weighted by n decision makers using linguistic variables: *VL—very low, L—low, M—medium, H—high* and *VH—very high.* Opinions of decision makers were then aggregated using the formula:

$$W_t = (1/n) \otimes (W_{t1} \oplus W_{t2} \oplus \dots W_{tn}), \quad t = 1, 2, \dots, k \tag{6.5}$$

where W_{tj} was the value of the weight for t-th criteria assigned by the j-th decision maker. A linguistic ranking scale used for describing the weights is shown in Fig. 6.2.

Next, the performance ratings for the above material characteristics and the cost for the given material are evaluated by the decision makers. This time the linguistic scale is: *worst* (W), *poor* (P), *fair* (F), *good* (G), and *best* (B). The membership functions are analogous to the one shown in Fig. 6.2.

Again the ratings of all decision-makers are aggregated:

$$R_{it} = (1/n) \otimes (R_{it1} \oplus R_{it2} \oplus \dots R_{itn}), \quad m = 1, 2, \dots, m, \; t = 1, 2, \dots, k \tag{6.6}$$

where R_{itj} was the aggregated rating of alternative i under criterion t assigned by the j-th decision maker.

After assignment of weights and ratings the final rating F_i for i-th alternative is evaluated using the formula:

$$F_i = (1/k) \otimes [(R_{i1} \otimes W_1) \oplus (R_{i2} \otimes W_2) \oplus \dots (R_{ik} \otimes W_k)] \tag{6.7}$$

where $F_i \approx (Y_i, Q_i, R_i, Z_i)$ is the approximated fuzzy number of the fuzzy suitability index of alternative i.

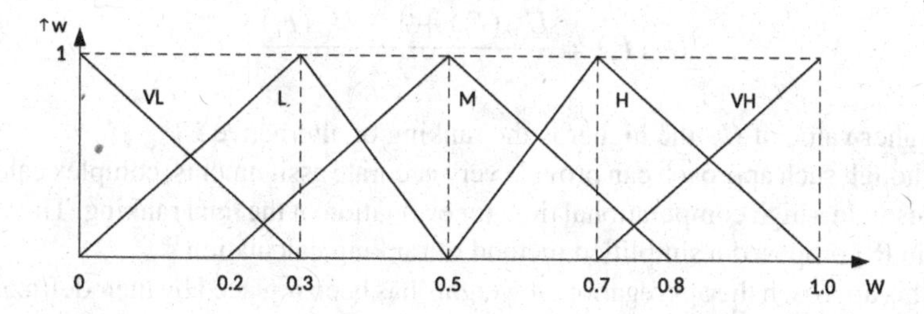

Fig. 6.2 Linguistic ranking scale for criteria weights [6]

The most suitable tool steel material for a given order in the proposed method is chosen after ranking such fuzzy ratings basing on simple maximisation and minimisation operators for fuzzy sets proposed by Chen [7]. Maximisation is defined as:

$$M = \{(x, f_M(x))|x \in \mathbb{R}\} \tag{6.8}$$

with the membership function defined as:

$$F_M(x) = \begin{cases} (x - x_1)/(x_1 - x_2), & x_1 \leqslant x \leqslant x_2, \\ 0, & \text{otherwise}. \end{cases} \tag{6.9}$$

Analogously, the minimisation is defined as:

$$G = \{(x, f_G(x))|x \in \mathbb{R}\} \tag{6.10}$$

with the membership function:

$$F_G(x) = \begin{cases} (x - x_2)/(x_2 - x_1), & x_1 \leqslant x \leqslant x_2, \\ 0, & \text{otherwise}, \end{cases} \tag{6.11}$$

where

$$\begin{aligned} x_1 = \inf D,\ x_2 = \sup D,\ D = \textstyle\bigcup_{i=1}^{m} D_i, \\ D_i = \{x | f_{F_i}(x) > 0\},\ i = 1, 2, \ldots, m \end{aligned} \tag{6.12}$$

The right utility function $U_M(F_i)$ and the left utility function $U_G(F_i)$ of each fuzzy rating evaluation F_i are defined, respectively, by:

$$U_M(F_i) = \sup_x (f_{F_i}(x) \wedge f_M(x)), \tag{6.13}$$

$$U_G(F_i) = \sup_x (f_{F_i}(x) \wedge f_G(x)). \tag{6.14}$$

The final value of ranking for alternative i is calculated as:

$$U_T(F_i) = \frac{U_M(F_i) + 1 - U_G(F_i)}{2} \tag{6.15}$$

The higher value of U_T the higher is the ranking of alternative i.

Although such approach can provide very accurate assignments, complex calculations result in a high computational time for evaluation of the final ranking. Therefore Chen in [8] proposed a simplified method of ranking calculation.

In this approach the aggregation of weights has been replaced by their deffuzification. A simple formula of deffuzification for trapezoidal fuzzy numbers is used [8]:

$$e = \frac{a + b + c + d}{2} \tag{6.16}$$

each criterion t assigned by j-th decision maker is defuzzified as:

$$w_{tj} = \frac{a_{tj} + b_{tj} + c_{tj} + d_{tj}}{4} \tag{6.17}$$

where $(a_{tj}, b_{tj}, c_{tj}, d_{tj})$ is a quadruple characterising linguistic variable W_{tj} represented by a fuzzy number.

The aggregated weight for each criterion t can be calculated as:

$$T(C_t) = \frac{w_{t1} + w_{t2} + \cdots + w_{tn}}{n} \tag{6.18}$$

Similarly the value of criterion t for potential candidate material A_i can be deffuzzified as:

$$r_{it} = \frac{p_{it} + x_{it} + y_{it} + z_{it}}{4} \tag{6.19}$$

where $(p_{it}, x_{it}, y_{it}, z_{it})$ is a quadruple characterising linguistic variable R_{it} represented by a fuzzy number.

The rating for material A_i based on the weighted criterion t can be achieved as a simple product of $T(C_i)*r_{it}$ except for the cost of the material. For this criterion a converting function is used:

$$F(A_i) = \frac{1}{v_i * (\frac{1}{v_1} + \frac{1}{v_2} + \cdots \frac{1}{v_m})} \tag{6.20}$$

Finally the rating for material A_i is calculated as:

$$R(A_i) = r_{i1} * T(C_1) + r_{i2} * T(C_2) + \cdots + r_{i(k-1)} * T(C_{k-1}) + F(A_i) * T(C_k) \tag{6.21}$$

Chen showed in a practical example that his method gave the same ranking as the method proposed by Wang and Chang [9].

6.2 Scheduling of Production Processes Using Fuzzy Logic

The most effective way to control continuous casting process is to run it as fast as possible, but with the respect to the metallurgical quality of the steel. The chemical and mechanical properties of the steel primary depend on the temperatures reached during the casting process. This process must be controlled in such a way that the temperatures fit in a specific range. Those temperatures are described by technologists in operation sheets according to the desired mechanical and chemical characteristic of the steel. The casting process can be controlled by regulating the speed of casting and cooling intensities for each coolant circuit. Necessary condition of the correct setting is that the material has to be solidified before it leaves from the caster. This property is so-called the metallurgical length—the length of liquid material from the meniscus [10].

In the planning and scheduling systems for steel production described in the literature uncertain parameters of the process and customers' requirements are principally expressed in the form of IF-THEN rules based fuzzy system.

Fuzzy logic can be seen as an extension of traditional Boolean logic based on the theory of fuzzy sets [11]. By introduction of the notion of degree in the verification of a condition, fuzzy logic enables a condition to be in a state other than only true or false (e.g. *partially true* that is between *completely true* and *completely false*). Thereby in many cases fuzzy logic may be more flexible for reasoning about phenomena that occur in real life, including the production processes that are not fully predictable and stable like in metal or chemical industry. Conventional logic forces the users to describe some inaccurate or uncertain parameters of the production process in a rough or approximate way, whereas sometimes it is easier and overall more accurate to express them in the form that is similar to the natural language (e.g., temperature is adequate).

The fuzzy system that is most commonly used in decision making consists of the set of rules in the following form:

$$\text{IF } \langle premise1 \rangle \text{ and } \langle premise2 \rangle \text{ and } \langle premiseN \rangle \text{ THEN } \langle conclusion \rangle$$

where $\langle premise1 \rangle$ is a statement of the type "x_i is of $L_{i,j}$", x_i represents a flexible predicate naming j-th linguistic term of the corresponding i-th linguistic variable and $L_{i,j}$ is given by a fuzzy set that represents the use of flexible predicate on the domain of x_i. The conclusion is also a fuzzy set representing a flexible predicate on the output of the system if the premises are satisfied [12].

Inference based on a fuzzy system and *modus ponens/modus tollens* rules requires the definition of a fuzzy equivalent to implication relationship $R(x, y) : A(x) \rightarrow B(y)$. Many definitions such relationships have been provided in the literature, but two of them are the ones most commonly used in practise:

- minimum operation: $R(x, y) = \min\{A(x), B(y)\}$ proposed by Mamdani [13]
- product operation: $R(x, y) = A(x) \cdot B(y)$ proposed by Larsen [14]

For compositional rule of inference min-max is used in Mamdani method and product-sum in Larsen.

Dorn et al. [15] proposed a reactive scheduling system for steel making process that uses fuzzy logic to create schedules that are robust with respect to changes due to certain types of event and fuzzy constraints to minimise the changes in the schedules, if rescheduling is required. The system has been built for a steel plant that produces high-grade steel from crude steel in a form of slabs and ingots. The products are then transported for subsequent processing like rolling and forging to the other plants. An outline of the steelmaking process in the steelworks is shown in Fig. 6.3.

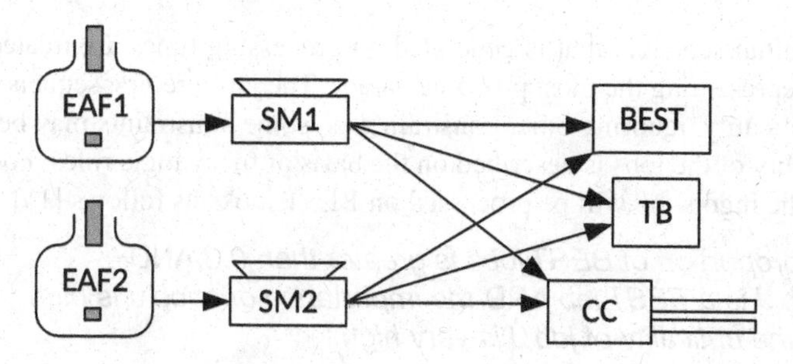

Fig. 6.3 The production flow in the steelworks analysed by Dorn [15]

Crude steel is melted two electric arc furnaces (EAF1 and EAF2), then it processed in the secondary metallurgy (SM1 and SM2) by pouring into a ladle furnace where fine alloying takes place, then in a vacuum oxygen decarburation unit. After that the steel is cast in a continuous caster (CC) to form slabs or into the moulds to form ingots. Solidification of ingots moulds requires a space in a teeming bay (TB) and for some large ingots, the Böhler-Electro-Slag-Topping (BEST) technology is used.

The authors enumerates three main kinds of fuzzy constraints (in the sense that they are not 100 % crisp) that must be taken into account during generation of schedules. First is the compatibility of the chemical properties of the steel between successive heats. As has been already described in the previous section, metal remaining after one heat in the electric arc furnace can interfere chemical characteristic of the next heat. In order to deal with this problem the engineers use as a rule of thumb that 3 % of a chemical element in a heat remain in the wall of the electric arc furnace, so 3 % of the difference of the elements in two consecutive heats should be assimilated by the second heat. The second group of constraints is related to due dates. Some steel forming processes require the cast to remain hot for subsequent treatments and for those castings (about 10 % of the production) due dates should be met within a tolerance of 2 h. Due to the nature of the casting process, it is also required the jobs on continuous caster to be scheduled either continuously or with sufficiently breaks enough for set-up or maintenance operations. The third group of constraints concerns capacity restriction. Continuous casting is a key process that directly influences production throughput and production costs, so it must be scheduled efficiently to reduce unnecessary breaks and reheatings. Contrary to it, the jobs for ingots that needs to solidify using BEST technology has to be scheduled with the certain time gaps between each other due to a long solidification time of such ingots.

The authors also postulate to describe processing times using fuzzy numbers with trapezoidal membership function as all processes have "an inescapable element of variability". Assuming that some jobs may take a little longer than average is especially important for electric furnaces, as the tardiness that occurs at this stage will propagate on the successive stages and finally cause that desired due dates are missing.

In the initial schedule that is generated all processing times are treated as crisp numbers, representing the most probable values. The jobs are first sequenced basing on their criticality, ignoring other constraints, so some constraints may be violated. The criticality of the jobs is described on the basis of fuzzy logic rules. For example a rule for the ingots the will be processed on BEST looks as follows [15]:

IF *the proportion of BEST-jobs is greater than 0.3* AND
a job J is a BEST-job AND *the importance of a job J is high*
THEN *the criticality of job J is very high*

The importance of the jobs is expressed by linguistic variables on the basis of the data reflecting the customers expectations and the market influence.

In the next step the schedule is evaluated for constraint violations. A schedule cannot be feasible if the degree of satisfaction of any constraint is below some critical threshold. The schedule is evaluated according to importance of the jobs and satisfaction of the constraints that have different weights:

$$\begin{aligned} f(S) &= J_i \in S \wedge importance\,(J_i) + \\ &\textstyle\sum_j \left(C_j \in S \wedge satisfaction\left(C_j\right) \cdot weight\left(type\left(C_j\right)\right)\right) \end{aligned} \tag{6.22}$$

In this step the crisp processing times and desired due dates are replaçed by fuzzy numbers.

Figure 6.4 shows that if the due date x is met within the interval of $[dd-1, dd+1]$, the membership function will evaluate to $A(x) = 1$, whereas $A(x) = 0.6$ for a due date violation of ± 2 h, which can be linguistically described as *early/late*, and finally it will evaluate to $A(x) = 0.3$ for a violation of ± 3 h, which can be described as *very early/very late*. Outside the fuzzy set it will evaluate to $A(x) = 0$ for a due date violation larger then ± 4 h. The membership function values in between can be linguistically described in terms of, e.g., *almost very good* for 0.7 and so on. In order to find fuzzy completion times for the jobs with assigned fuzzy due dates (e.g. for the BEST unit) processing times of all the jobs that precede such jobs must be fuzzified. Exemplary definitions of the processing times for electric furnaces, secondary metallurgy, continuous caster and finally the BEST unit are shown in Fig. 6.5.

Despite the final schedule is generated in a fairly precautionary manner, some disturbances in the production process cannot be envisaged, for example if some job took longer to complete that it was expected, what may result in due date violation. Such tread can be evaluated and, if necessary, a repair mechanism can be applied.

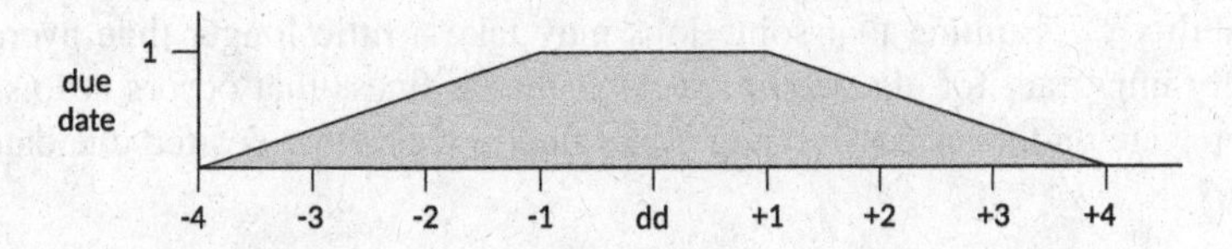

Fig. 6.4 Membership function evaluation for a due date

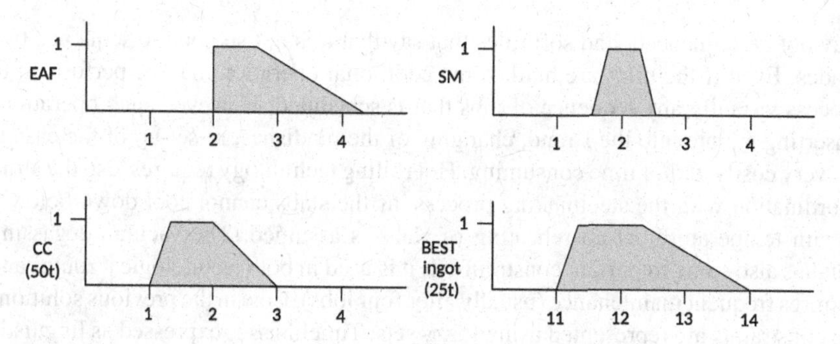

Fig. 6.5 Fuzzy definition of processing times on different production units

Tabu search algorithm is used to find which jobs should be exchanged to repair the schedule. In order to limit the number of necessary re-scheduling thresholds are defined to measure significance of adverse events and to assess their impact on the final schedule evaluation (i.e. whether some constraints may remain violated).

More complex production process system of the highly specialised steel in the LD3 plant in Linz was analysed by Dorn in [16]. The overall flow of the steel production is shown in Fig. 6.6. The blast furnace (BF) delivers the pig iron either to the desulfurization unit (DS), to the mixer (MX), or directly to one of the three converters (CV7, CV8, CV9), however only two of them are usually used. The mixer is a 2000t container that plays a role of a stock between BF and the converters. There are two typical production routes: the steel from CV7 is poured into a ladle and delivered to the ladle furnace (LF) and later to the single-stranded continuous caster CC3, and the steel from CV9 is delivered to the conditioning stand (CS) and to the two-stranded caster CC4. On both routes also the vacuum-degassing unit (VAC) can be involved. About 90 % of the cast slabs are delivered either hot to the hot rolling mill (HRM) or stored in the slab stock (SS). The remaining jobs are produced for the forge and the foundry [16].

Like in the previous studies, the steel grade compatibility constitutes the main group of constraints for the blast furnace and continues casters. The author noticed, that there are two kind of such constraints: hard constraints saying that two grades

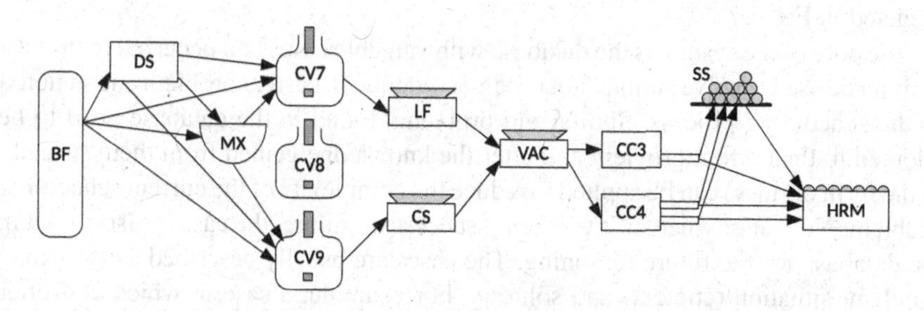

Fig. 6.6 Steel production process analysed by Dorn in Linz LD3 steelworks [16]

may not be sequenced, and soft rules that say that it is not so good to sequence two grades. Even if the rules are hard, some additional operation may be performed to process virtually any sequence of jobs that is scheduled. However, such operations (inserting a plate into the strand, changing of the tundishes, re-set-up of the caster) are very costly and/or time consuming. Hot rolling technology requires also the strict coordination with the steelmaking process, as the slabs cannot cool down below a certain temperature, as no reheating of slabs is assumed. The vacuum-degassing unit has also some important constrains, as it is used in both technological routes and requires frequent maintenance (usually after four jobs). Like in the previous solutions the constraints are represented using fuzzy sets. Timeliness is expressed as linguistic variables and takes values like *very early*, *in time*, etc. The same is with the quality description (e.g. *bad*, *very high*), lower limit (e.g. *zero*, *few*, *too many*), amount (e.g. *few*, *sufficient*, *too much*), priority (e.g. *normal*, *very important*) and the sequence length on the casters (e.g. *very short*, *too long*). Some values of such variables are not allowed in a valid schedule if the constraint is treated as a hard one. For example if a group casting jobs last *too long* this violates the necessary maintenance of the tundish and cannot be accepted. In order to better deal with soft constraints when searching for a valid schedule Dorn introduced weights for each type of constraints. For example the most costly set-up of continuous caster has weight of 1.0, while due dates or compatibility violations are weighted as 0.6 and undesirable format changes only as 0.2.

The schedule is evaluated on the basis of weighted constraints satisfaction and importance of jobs, which is also expressed by fuzzy sets (similar to the system that has been described in the beginning of this section). The finale value of schedule estimation is always between 0 and 1, where 1 means the optimal value. Schedules can be rated as good if the evaluation is above 0.9.

If the schedule needs to be repaired due to some unpredicted disturbances in the production process, again Dorn suggest using tabu search algorithm and repair heuristic as he found them a very effective one if the problem is not too complex. However, if the constraints cannot be easily satisfied due to more restrictive technological and organisational dependencies and/or more complex production routes (in this case taking into account three converters instead of two) the performance of the system may significantly decrease [16]. To face this problem Dorn propose to take advantage of a case-based reasoning system. The overview of such system is depicted in Fig. 6.7.

The core of the system is the database with valuable cases that occurred in the past and can be used to solve similar, not necessary identical, current problem encountered in the scheduling process. Similar old problems found in the database need to be adopted to the current problem and after the knowledge gained from them (usually in the form of rules) can be applied to reduce the complexity of the current scheduling problem. No matter whether it will bring success or failure, the case is also stored in the database for the future reasoning. The cases are usually described as the triple: problem, situation (context) and solution. For example, a case in which converter CV7 was replaced by the converter CV8 was described as:

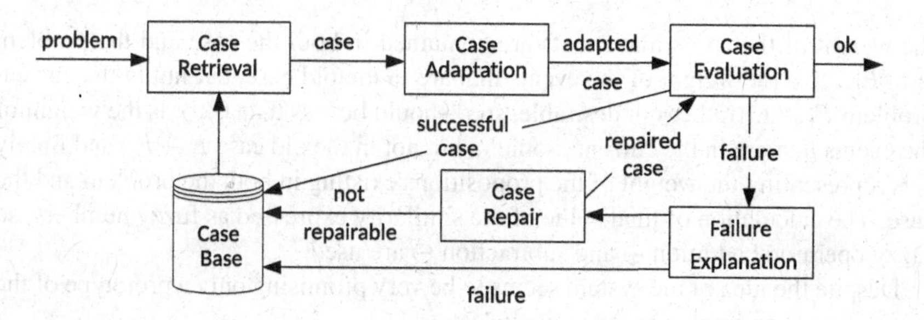

Fig. 6.7 Case reasoning system for fuzzy reactive scheduling [16]

problem: *break down of CV7 for some hours*
situation: *many jobs on CV7, free CV8*
solution: *exchange CV7 with CV8 in schedule*

Note that in description of the cases fuzzy sets are used to soften the constraints. This can be better seen in the description of the following case respecting the shortage of pig iron:

problem: *very few pig iron*
situation: *medium set-ups on CC3, few set-ups on CC4,*
many quality separations on CC3, medium quality separations on CC4,
few tundish changes on CC3, medium tundish changes on CC4,
some important jobs on CC4
solution: *increase weight set-ups CC3, increase weight important jobs*
strong decrease weight set-ups CC4, optimize schedule

The above case shows also an interesting approach to solve the problem. Instead of finding a new sequence of jobs on casters that requires less steel that may be very time consuming, only the weights for those casters are updated and the schedule is re-optimised.

A crucial stage in the inference process is to find cases that are similar to the current problem, as it is necessary to use some mechanism to express and measure the similarity. If a problem is described by a set of events E and a set of propositions P, E_i and P_i represent the set of events and the set of propositions for the case C_i, the following formula is used to determine the similarity of the problem to the case:

$$\begin{aligned} S(C_i) = \alpha & \sum_{\forall e_j \in (E \cap E_i)} (1 - |\delta(e_j, E_i) \Theta \delta(e_j, E)|) \\ \oplus \beta & \sum_{\forall e_j \in (E_i - E)} (1 - \delta(e_j, E)) \oplus \chi \sum_{\forall e_j \in (E - E_i)} (1 - \delta(e_j, E_i)) \\ \oplus \gamma & \sum_{\forall p_j \in P_i} (1 - |\delta(p_j, P_i) \Theta \delta(p_j, P)|) \end{aligned} \tag{6.23}$$

where δ is the function that tells if the event or proposition j exists in the current problem or the case (1 for crisp relationships, fuzzy number, otherwise), α is the weight of the cases that events are contained in both the case and the problem $E \cap E_i$, β is the weight of the events that are in the old case, but not in the current problem $E_i - E$ (rather not desirable, so β should be less than 0), χ is the weight of the events that are in the current problem, but not in the old case $E - E_i$, and finally γ is representing the weight of the propositions existing in both the problem and the case. The calculation of final value of the similarity expressed as fuzzy numbers, so fuzzy operators: addition $\oplus$ and subtraction $\ominus$ are used.

Despite the idea of the system seems to be very promising only a prototype of the system has been implemented in practise.

Another system basing on knowledge base and fuzzy logic was proposed by Adenso-Diaz et al. [9]. The proposed system was dealing with planning in a roll shop and was implemented in one of the Spanish steelworks. Roll shops serve for storing and maintenance of rolls used in hot rolling and cold rolling mills. Rolls depend on their planned location in the mill (type of the mill, a stand in the mill) and the type of the product that can be rolled using them. There was about 150 different roll types in the analysed steelworks. The authors developed a fuzzy system to plan the work in the roll shop by determining the priority of the rolls. The system contains over 50 variables that can be divided into following four groups:

- Frequency of incidents: An undetermined number of rolls suffer certain processing problems and never get to the stand; other rolls have a lamination time lower than that accepted as normal.
- Mill/Stand importance: Some stands or mills may be more conflictive or key than others, due to the frequency of incidents or the importance/urgency of the orders.
- Estimation deviation: Deviation between the estimation of rolls to be sent to the mills and the actual rolls sent.
- Optimum deviation: The most important variable for priority evaluation. This is the roll shop time needed to get the optimum number of rolls with the current use of resources.

The proposed system has two main modules. The goal of the first is to gather data from the database and to transform it into the four types of inputs presented above. The second is the fuzzy system itself, which provides priority of processing in the roll shop for every roll type. The inputs are fuzzified and 39 fuzzy rules are used to produce the final output that after centroid defuzzification is presented as a crisp number. In membership functions linguistic variables are used. Exemplary rule looks as follows

IF *Optimum_deviation is medium* AND *Mill/stand_importance is high*
THEN *Priority is high*

The system has been implemented in practise and, according to the authors, three main benefits from its application can be enumerated:

- decrease in costs, mainly stock and warehousing costs,
- increase in security, because the system attenuates the information handicap,
- general improvement in quality.

6.3 Cooperation of Planning with Fuzzy Logic

Dorn and Kerr [17] proposed to apply cooperative scheduling to optimise schedules in a steelmaking plant and a rolling mill for VA Stahl steelworks in Linz (Austria). In this approach fuzzy logic was used to perform so called fuzzy communication in order to agree two separate schedules in different plants that may have different technological and organisational constraints. As it was already mentioned, the main group of constraints for the steelmaking plant is related to the compatibility: the steel grade of consecutive heats must be similar, castings width must stay within a given range and the degassing procedure in the secondary metallurgy must be consistent. Also for the rolling mill, subsequent jobs must have similar steel quality, but different constraints on width and thickness of rolled goods must be considered. Another problem is that not all steel qualities can be rolled warm, so some slabs must cool down before they can be processed in the rolling mill.

Scheduling process works in the proposed system works in such a way that the schedule generated for one plant is evaluated and sent to the second plant. Instead of firm constraints, fuzzy sets and so called fuzzy sequences are used. For example the schedule generated for the rolling mill consists of a list of jobs, a number of fuzzy sets describing windows, when casting can be performed, and a set of fuzzy sequence constraints that constrain the sequence of jobs. Figure 6.8 shows that job j_1 should be done first, job j_{12} should be rolled in the end, as it has the lowest weight and job j_2 should be performed sometime between job j_1 and j_{12}.

The subsystem in the steelmaking plant adds its own constraints, evaluates the schedule and repairs it to achieve better evaluation using a tabu search procedure. After that, the schedule is sent back to the rolling mill plant. If the evaluation of the new schedule is worse than before, the subsystem in the rolling mill plant tries to repair the schedule. A repaired version of the schedule is sent again to the steelmaking plant and if its evaluation is better than it was earlier, it is accepted, otherwise it is rejected.

According to Dorn despite that the proposed system was not tested in practise, fuzzy communication process enables for sharing more information than in the traditional crisp data communication. The prototype of the system itself has not been built according to multi-agent architecture, and communication is done sequentially, what may limit the potential benefits from application of fuzzy sets and fuzzy systems for production scheduling of several plants working in a common supply chain.

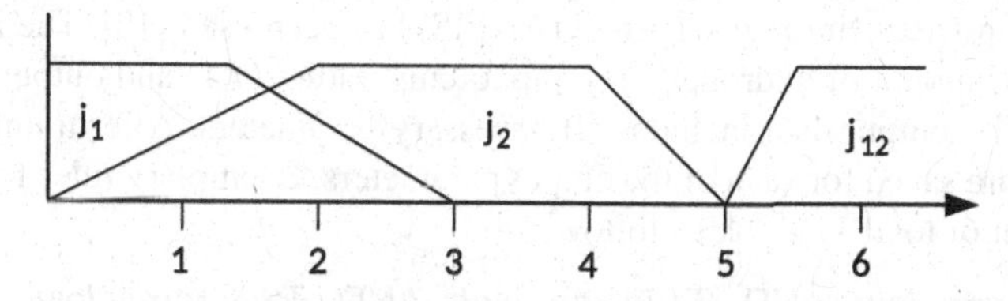

Fig. 6.8 Fuzzy sequence of rolling mill jobs

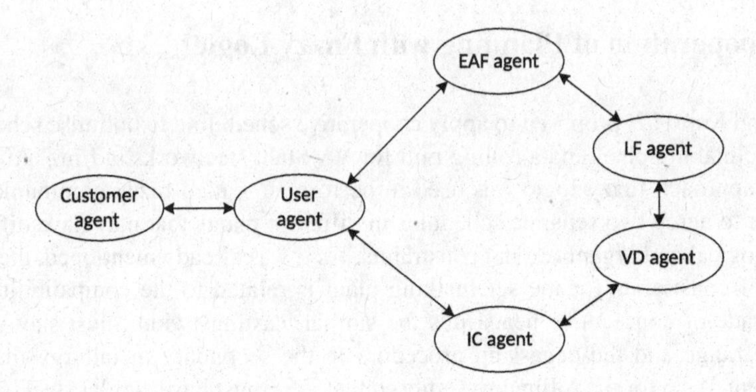

Fig. 6.9 A fuzzy multi-agent system the planning and scheduling in steelworks

Fazel Zarandi and Ahmadpour [2] proposed a much more complex solution for the cooperative planning and scheduling in the steel industry. Their system is based on multi-agent architecture and adaptive neuro-fuzzy networks. An overview of the system is shown in Fig. 6.9.

The system consists of six agents. Agents are built around several modules: user interface, communication interface, reasoning module that uses knowledge base and learning module. Customer agent is responsible transforming customers' orders into production orders after negotiations with a User agent. The User agent deals with customers' orders and communicates with other agents to process them. The role of the Ingot Casting (IC) agent is to supply ingots produced in ingot moulds to the User agent. The Vacuum Degassing (VD) agent is responsible for determining parameters of degassing process. The Ladle Furnace (LF) agent is responsible for determining parameters of molten steel refining process according to the customer's order. Finally, the Electric Arc Furnace (EAF) agent supplies molten steel to the LF agent.

The User agent has to assess the order sent from the Customer agent whether it is a feasible one, taking into account cost of necessary additives and cost of delivery date differences. If the order receives a status of "infeasible" status the User and the Customer agents may negotiate different delivery date and/or required properties. The IC agent determines the amount of molten steel that is necessary to produce the ingots ordered by the customer and provides the estimation of total casting processing time for the order. Both VD and LF agents have to deal with uncertain parameters of the production process. The VD agent must determine such parameters as vacuum pressure, the amount of neutral gas (Argon) and processing time. For this purpose, an adaptive neuro-fuzzy inference system (ANFIS) has been used [18]. The input data includes the amount of hydrogen (H), input temperature ($T1$) and output temperature ($T2$). The output data includes all necessary parameters. After training stage, fuzzy rules are saved for each of the output parameters. Exemplary rules for vacuum pressure (out of total 11) look as follows:

IF *H is very low* AND *T1 is very high* AND *T2 is rarely low*
THEN *Pv is* $(-0.1848 * H - 0.6485 * T1 + 1.8305 * T2 - 178.1581)$

IF *H is rarely low* AND *T1 is very strongly low* AND *T2 is moderate* THEN *Pv is* $(3.1852 * H - 4.1816 * T1 + 4.2968 * T2 + 7.7689)$

IF *H is rarely high* AND *T1 is moderate* AND *T2 is very high* THEN *Pv is* $(0.5049 * H + 0.9166 * T1 + 0.4582 * T2 - 78.2941)$

After learning phase total 54 rules have been stored in the knowledge base.

The Ladle Furnace agent must determine such parameters of the refining process as the amounts of additives and the processing time. Also in this case ANFIS system is used to collect rules in the knowledge base. The input data includes the chemical properties of input molten steel (S2, Cr2 and Ni2), temperature and required chemical properties of output molten steel. The output data includes the amounts of needed additives to change the chemical properties of molten steel, and processing time. Exemplary rules for processing time (out of 12) are shown below:

IF *S2 is rarely low* AND *Cr2 is moderate* AND *Ni2 is high* AND *Temperature is rarely high* THEN *Time is* $(0.5138 * S2 - 0.0357 * Cr2 - 1.2426 * Ni2 + 0.0843 * Temperature - 106.6996)$

IF *S2 is moderate* AND *Cr2 is very high* AND *Ni2 is rarely low* AND *Temperature is rarely low* THEN *Time is* $(0.1389 * S2 + 1.6269 * Cr2 - 1.4204 * Ni2 + 2.4173 * Temperature - 4145.3457)$

IF *S2 is low* AND *Cr2 is very low* AND *Ni2 is strongly high* AND *Temperature is very low* THEN *Time is* $(-13.0052 * S2 + 61.592 * Cr2 - 296.1599 * Ni2 - 0.7135 * Temperature + 307.2859)$

This time 111 rules have been generated after learning phase.

The system was implemented in MATLAB7. A single agent was used for every process, however in real application more agents can work for a single process. The authors have not provided any information about the plans on implementation of the proposed system in real steelworks.

Later Zarandi and Azad [18] presented a multi-agent system based on Type-2 fuzzy sets. The authors enumerate four sources of uncertainty about membership functions used in classical fuzzy sets:

- different meaning of the same word to different people,
- experts opinions may have histogram of values, especially if they vary,
- measurements activating a type 1 fuzzy logic system may be noisy,
- the data used to tune the parameters of a type 1 fuzzy logic system may also be noisy.

In the Type-2 fuzzy logic system rules look as follows [19]:

IF x_1 *is of type* $\tilde{F}_1$ AND ... AND x_p *is of type* $\tilde{F}_p$ THEN *y is of type* $\tilde{Y}$

The scheduling system was built using a combination of fuzzy programming and fuzzy contract net protocol proposed by Li et al. in [20]. In such approach an Order agent transfers an order to a Scheduling agent and when the order arrives, the Scheduling agent models a scheduling programming problem and then selects an algorithm to solve the problem (e.g., some heuristic like genetic algorithm). The Scheduling agent must also adjust the initial schedule to a new condition that occurred during the production process, like machine breakdown or appearance of a new task. In the fuzzy programing stage a Type-2 fuzzy flow shop problem formulation is used to find the sequence of the n jobs maximising the total satisfactory degree. Type-2 fuzzy due date is represented by the degree of satisfaction with respect to the completion time c_j, and denoted by a doublet:

$$\tilde{D}_j(d_j^L, d_j^U) \tag{6.24}$$

where d_j^U is not fixed, since penalties for the delay of the jobs are defined differently by the experts as:

$$d_j^U = d_j^L + \rho_k \tag{6.25}$$

in which penalty

$$\rho_k \in [\rho_{k_1}, \rho_{k_2}].$$

Figure 6.10 shows the fuzzy due date and the membership function of satisfactory degree $\mu_j(c_j)$. The objective function is then defined as maximising total satisfactory:

$$\max f_{sum} = \sum\nolimits_{j=1}^{n} \mu_j^*(c_j) \tag{6.26}$$

where μ_j^* is the final crisp value of the interval set of a due date set calculated as the average of upper and lower bound.

When an emergency event occurs fuzzy, a contract protocol is used. First, the Scheduling agent builds some Task agents, each for one or more jobs. The Task agent announces the job (or group of jobs) to the Resource agents inviting them to bide for it. After receiving the answers, the Task agent selects the best Resource agent by evaluating the received bids and it is removed from the system [20]. Each agent has a fuzzy module that includes fuzzifier, rule base, fuzzy inference engine, and output processor (type-reducer and defuzzifier). A detailed structure of the fuzzy module is shown in Fig. 6.11.

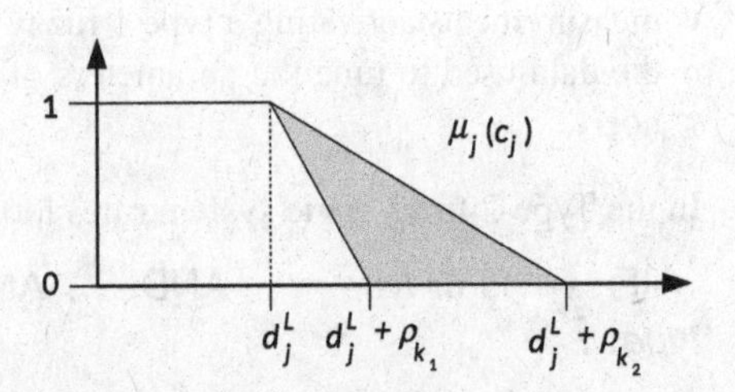

Fig. 6.10 Fuzzy due date and the membership function of satisfactory degree

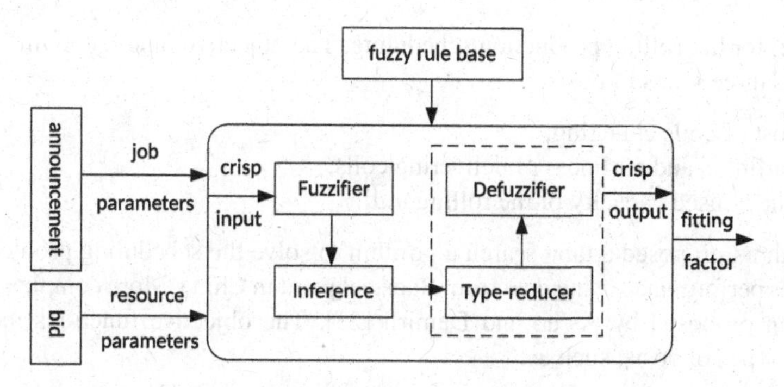

Fig. 6.11 Fuzzy due date and the membership function of satisfactory degree

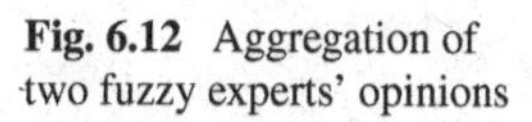
Fig. 6.12 Aggregation of two fuzzy experts' opinions

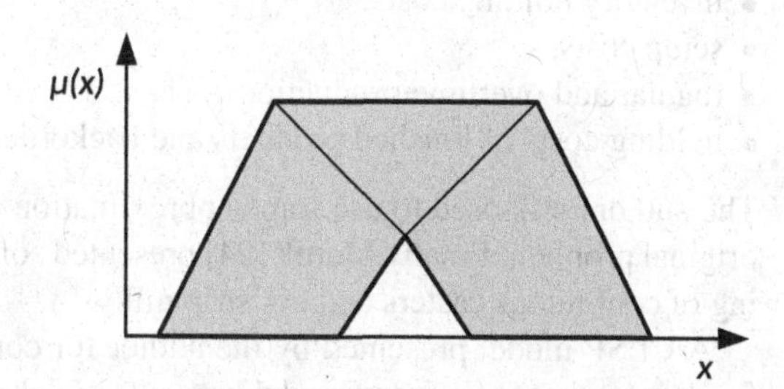

The fuzzifier changes the crisp values of the parameters into three linguistic variables (e.g., for completion time: *early, medium, late*; for priority: *high, medium, low*; for due date: *long, medium, short*). Those linguistic variables can be defined as Type-2 membership function, shape of which is obtained basing on experts' opinions, which are written as Type-1 fuzzy numbers. Aggregation of two experts' opinions into Type-2 membership function is shown in Fig. 6.12.

The contract fuzzy protocol uses fuzzy rules. For example in announcement stage following rules are used:

IF *due date is short* THEN *priority of announcement is high*
IF *due date is medium* THEN *priority of announcement is medium*
IF *due date is long* THEN *priority of announcement is low*

6.4 Continuous Caster Scheduling Based on Fuzzy Lot-Sizing

A scheduling agent that uses a fuzzy version of a capacitated lot-sizing problem (CLSP) is presented in this section. Mathematical integer programing (MIP) models basing on capacitated lot-sizing model and its variations [21] is one of the approaches used for planning and scheduling in steel industry. Wang and Tang [22] presented a model for hot rolling production scheduling. The objective function in this model includes three kinds of costs:

- the cost of coils changing,
- the earliness and tardiness of delivering coils,
- and the unused capacity of the rolling mill.

The authors proposed a tabu search algorithm to solve the scheduling problem and tested its performance using data from Baosteel plant in China. More complex model has been proposed by As'ad and Demirli [23]. The objective functions includes various types of costs, such as:

- ordering and purchasing of raw materials,
- inventory holding costs,
- setup costs,
- regular and overtime production costs,
- holding costs of finished products and backorder costs.

The authors proposed to use some approximations schemes in order to simplify the original problem. Finally Mattik [24] presented some models for integrated scheduling of continuous casters and hot strip mills.

A CLSP model presented by the author for continuous casting can be the basis for development of a fuzzy model used by the Scheduling agent. A modified version of the model is defined as follows:

$$\min \tilde{Z}_{it} = \lambda_1 \sum_{m\in M} \sum_{t\in T} (\tilde{I}^{c-}_{mt} + \tilde{I}^{c+}_{mt}) + \lambda_2 \sum_{l\in L} \sum_{m\in M|l\neq m} \sum_{t\in T} st^c_{lm} x^c_{lmt} + \lambda_3 \sum_{m\in M} (\sum_{t\in T} cs^c_m Z^c_{mt} / C^c_t) \tag{6.27}$$

constraints:

$$\tilde{I}^{c+}_{m,t-1} - \tilde{I}^{c-}_{m,t-1} + Z^c_{mt} - \tilde{I}^{c+}_{mt} + \tilde{I}^{c-}_{mt} = \tilde{d}^c_{mt}, \quad m \in M,\ t \in T \tag{6.28}$$

$$\sum_{m\in M} Z^c_{mt} cs_m + \sum_{l\in L} \sum_{m\in M|l\neq m} st^c_{lm} x^c_{lmt} \leq C^c_t, \quad t \in T \tag{6.29}$$

$$cs^c_m Z^c_{mt} \leq C^c_t y^c_{mt}, \quad m \in M, t \in T \tag{6.30}$$

$$x^c_{lmt} \geq y^c_{l,t-1} + y^c_{mt} - 1, \quad l \in M, m \in M|l \neq m, t \in T \tag{6.31}$$

$$\sum_{m\in M} y^c_{mt} = 1, \quad t \in T \tag{6.32}$$

$$Z^c_{mt} \geq ml^c y^c_{mt}, \quad m \in M, t \in T \tag{6.33}$$

$$x^c_{lmt} \in \{0, 1\}, \quad l \in M, m \in M | l \neq m, t \in T \tag{6.34}$$

$$y^c_{mt} \in \{0, 1\}, \quad m \in M, t \in T \tag{6.35}$$

$$Z^c_{mt} \geq 0 \quad m \in M, t \in T \tag{6.36}$$

Indices/sets:

$t \in T$ time period;
$m \in M$ steel grades;

Parameters:

$\tilde{d}_{mt}$ demand for steel grade m in period t;
cs_i capacity consumption for item I;
$g_{ik} = 1$ if item i is produced from steel grade k, otherwise 0;
st_k setup cost for steel grade k;
C_t capacity a heat in period t;

Variables:

$\tilde{I}^+_{it}, \tilde{I}^-_{it}$ items i delayed (–) and stored (+) at the end of period t;
$z_{tk} = 1$ if there is a setup (resulting from a change in grade) in period t;
$y_{tk} = 1$ if steel grade k is produced in period$t$, otherwise 0;
x_{it} number of items i produced in period t.

It should be noted that the above model does not include many technological constraints such as, e.g., adequate slabs width in a sequence or their temperature. Such model should then be considered as a first stage of the continuous casting production planning process, in which the quantities of particular steel grades are determined. In the seconds stage jobs on continuous casters has to be sequenced with respect to the technological constraints. As it was shown in the previous sections, such constraints may also include some fuzzy rules, due to uncertainty of the casting process. The same concerns the steel grade, despite the model assumes predetermined steel grades for the slabs, this can be easily change with the use of fuzzy logic, saying which steel grade maybe assigned to satisfy customers' requirements.

Lot-sizing with fuzzy parameters is rather rarely studied in the literature, especially when compared to fuzzy shop scheduling problems. Yan et al. [25] analysed lot-sizing production planning problem with profits, customer demands and production capacity characterised by fuzzy variables with trapezoidal membership functions. To solve the problem the authors proposed to apply a standard genetic algorithm performing fuzzy simulation. Rezaei and Davoodi [26] studied a lot-sizing problem with supplier selection under fuzzy demand and costs (price, transaction cost and holding cost) with triangular membership functions. Also in this case a standard genetic algorithm was used to determine upper and lower bound for production quantities.

Most recently Sahebjamnia and Torabi [27] considered a multi-level capacitated lot-sizing problem with uncertain setup, holding, and backorder costs expressed as a fuzzy numbers with trapezoidal membership functions. They proposed a heuristic in which uncertain constraints as well as imprecise objective functions are converted into the crisp values by using the expected interval and value of the ill-known parameters, respectively. Then the authors solved such a problem using a standard MIP solver.

When planning continuous casting process, the demand for a given period of time is usually taken as a strict constraint. However strict deadlines are necessary mainly for the cases when a minimal temperature of slabs is required in further processing (hot rolling). It is especially true for direct hot charge and hot charge rolling technologies (see [1] for the classification of hot rolling technologies). On the other hand the demand for a given steel grade depends on the delivery dates that were confirmed to the customer. Thus in our solution we introduced rules telling if the desired term for steel grade (demand in a give period) may vary within some limits. The exemplary rules are as follows:

IF *rolling temperature constraint is obligatory* THEN *demand for grade is strict*

IF *due date is short* THEN *demand for grade is strict*

IF *rolling temperature constraint is strong* THEN *demand for grade is tight*

IF *due date is medium* THEN *demand for grade is tight*

IF *rolling temperature constraint is weak* THEN *demand for grade is normal*

IF *due date is long* THEN *demand for grade is normal*

A linguistic variable *demand for grade* is mapped for a fuzzy demand with a triangular membership function. For the *strict* demand membership function is defined as: (d_{mt}*0.99, d_{mt}, 0), for *tight demand* it is defined as: (d_{mt}*0.95, d_{mt}, d_{mt}*1.05), and for *normal* demand (d_{mt}*0.9, d_{mt}, d_{mt}*1.1).

To solve the fuzzy capacitated lot-sizing problem for continuous casting scheduling a dedicated genetic algorithm was used. Contrary to the problems presented in the literature, in the computational experiment we consider a scheduling problem of an industrial-size, so a standard genetic algorithm cannot be applied effectively to receive acceptable solutions in a reasonable time. Instead, we applied a genetic algorithm with two dedicated repair algorithms, described in [28] and adopted to the fuzzy scheduling of continuous caster. A chromosome representing the solution consists of the values of x_{it} variables. A non-standard crossover operator, copying one period from a parent's chromosome to a child's chromosome, and irregular mutation [29] were used as recombination operators.

The following parameters of GA where used:

- crossover rate $c_R = 0.8$,
- mutation rate $ml_R = 0.05$,
- population size $pop_{size} = 50$ solutions.

The algorithm has been tested on a problem that includes 50 different products, 10 different steel grades and a time period with 48 discrete time slots (hours). The minimum size of a cast was set to 10 ton. The other parameters have been generated from a uniform distribution:

- demand for steel in period $t - d_{mt} \in [100, 300]$ tons,
- capacity consumption $cs_{mt} \in [0.1, 0.3]$ min/ton,
- setup time for changing steel grade $st_{lm} \in [30, 60]$ min.

As it has been already explained, the demand for the particular steel grade is fuzzified according to the values of linguistic variables in two variants. In the *tight* variant 30 % of the steel grades had a *strict* demand, 30 % had a *tight* demand and the remaining ones had *normal* demand. In the *loose* variant 5 % of the steel grades had a *strict* demand, 25 % had a *tight* demand and the remaining ones had *normal* demand.

The genetic algorithm has been run for 20 times for the same variant of the demand and the average results for the best solution in the population after 20,000 generations (ca. 1 min for a single run) have been collected. The results are shown in Table 6.1.

The capacity utilisation of continuous caster is provided together with the production costs, including the holding costs and the penalty for tardiness and setup costs. As it was expected, loosening the demand constraints brings a significant decrease in production costs (up to 50 %), but also a noticeable increase in capacity utilisation (up to 12 %). Even though in the real production systems such advantages cannot be achieved due to technological and organisational constraints that have been omitted in the proposed lot-sizing model, it has been shown that the application of fuzzy set theory to planning and scheduling of steel production processes may bring significant savings.

Table 6.1 The average values of capacity utilisation and cost function for different variants of CC scheduling

	Capacity utilisation [%]	Cost function [penalty points]
Crisp scheduling	0.75	14694
Fuzzy scheduling with thigh variant	0.81	10390
Fuzzy scheduling with loose variant	0.84	9833

References

1. Tang, L., J. Liu, A. Rong, and Z. Yang. 2001. A review of planning and scheduling systems and methods for integrated steel production. *European Journal of Operational Research* 133(1): 1–20.
2. Zarandi, M.H.F., and P. Ahmadpour. 2009. Fuzzy agent-based expert system for steel making process. *Expert Systems with Applications: An International Journal* 36(5): 9539–9547.
3. Dorn, J. 1996. Expert systems in the steel industry. *IEEE Expert* 11(1): 18–23.
4. Vasko, F.J., K.L. Stott, F.E. Wolf, and L.R. Woodyatt. 1989. A fuzzy approach to optimal metallurgical grade assignment. In *Applications of fuzzy set methodologies in industrial engineering*, ed. G.W. Evans, W. Karwowski, and M.R. Wilhelm, 285–298., Advances in Industrial Engineering Amsterdam: Elsevier Science Publishing.
5. Woodyatt, L.R., K. Stott, F.E. Wolf, and F.J. Vasko. 1992. Using fuzzy sets to assign metallurgical grades to steel. *JOM, The Journal of The Minerals, Metals & Materials Society (TMS)* 44(2): 28–31.
6. Wang, M.-J.J., and T.-C. Chang. 1995. Tool steel materials selection under fuzzy environment. *Fuzzy Sets and Systems* 72(3): 263–270.
7. Chen, S.-H. 1985. Ranking fuzzy numbers with maximizing set and minimizing set. *Fuzzy Sets and Systems* 17(2): 113–129.
8. Chen, S.-M. 1997. A new method for tool steel materials selection under fuzzy environment. *Fuzzy Sets and Systems* 92(3): 265–274.
9. Adenso-Díaz, B., I. González, and J. Tuya. 2004. Incorporating fuzzy approaches for production planning in complex industrial environments: The roll shop case. *Engineering Applications of Artificial Intelligence* 17(1): 73–81.
10. Mauder T., J. Štětina, and M. Masarik. 2013. On-line fuzzy regulator for continuous casting process. In *In:Proceedings of 22nd International Conference on Metallurgy and Materials METAL*, 23–38, Brno, Czech Republic, 15–17 May 2013.
11. Zengchang, Qin, and Yongchuan, Tang. 2014. *Uncertainty modeling for data mining: A label semantics approach*. Springer.
12. Moraga, C. 2005. Introduction to fuzzy logic. *Facta universitatis - series: Electronics and Energetics* 18(2): 319–328.
13. Mamdani, E.H., and S. Assilian. 1975. An experiment in linguistic synthesis with a fuzzy logic controller. *International Journal of Man-Machine Studies* 7(1): 1–13.
14. Larsen, M.P. 1980. Industrial applications of fuzzy logic control. *International Journal of Man-Machine Studies* 12(1): 3–10.
15. Dorn, J., R.M. Kerr, and G. Thalhammer. 1995. Reactive scheduling: Improving the robustness of schedules and restricting the effects of shop floor disturbances by fuzzy reasoning. *International Journal of Human-Computer Studies* 42(6): 687–704.
16. Dorn, J. 1995. Case-based reactive scheduling. In *Artificial intelligence in reactive scheduling*, ed. E. Szelke, and R.M. Kerr, 32–50. Kluwer Academic Publishers.
17. Dorn, J. and R.M. Kerr 1994. Co-operating scheduling systems communicating through fuzzy sets. In *Preprints of the 2nd IFAC/IFIP/IFORS-Workshop on Intelligent Manufacturing Systems IMS94*, 367–373, Vienna.
18. Zarandi, F.M.H., and K.F. Azad. 2013. A type 2 fuzzy multi agent based system for scheduling of steel production. In *IEEE Joint IFSA World Congress and NAFIPS Annual Meeting, IFSA/NAFIPS*, 992–996, Edmonton, Alberta, Canada, 24–28 June 2013.
19. Castillo, O., and P. Melin. 2008. *5 Design of Intelligent Systems with Interval Type-2 Fuzzy Logic*, vol. 223, Studies in Fuzziness and Soft Computing Berlin: Springer.
20. Li, Y., J.-Q. Zheng, and S.-L. Yang. 2010. Multi-agent-based fuzzy scheduling for shop floor. *The International Journal of Advanced Manufacturing Technology* 49(5–8): 689–695.
21. Helber, S. 1995. Lot sizing in capacitated production planning and control systems. *Operations-Research-Spektrum*, 17(1): 5–18.

22. Wang, X., and L. Tang. 2008. Integration of batching and scheduling for hot rolling production in the steel industry. *The International Journal of Advanced Manufacturing Technology* 36(5–6): 431–441.
23. Asad, R., and K. Demirli. 2010. Production scheduling in steel rolling mills with demand substitution: Rolling horizon implementation and approximations. *International Journal of Production Economics* 126(2): 361–369.
24. Mattik, I. 2014. *Integrated scheduling of continuous casters and hot strip mills: A block planning application for the steel industry*. Produktion und Logistik: Gabler Verlag.
25. Yan, W., J. Zhao, and Z. Cao. 2005. Fuzzy programming model for lot sizing production planning problem. In *Fuzzy systems and knowledge discovery*, ed. L. Wang, and Y. Jin, 285–294, Lecture Notes in Computer Science Berlin: Springer.
26. Rezaei, J., and M. Davoodi. 2006. Genetic algorithm for inventory lot-sizing with supplier selection under fuzzy demand and costs. In *Advances in applied artificial intelligence*, ed. M. Ali, and R. Dapoigny, 1100–1110, Lecture Notes in Computer Science Berlin: Springer.
27. Sahebjamnia, N., and S.A. Torabi. 2014. A fuzzy stochastic programming approach for multi-level capacitated lot-sizing problem under uncertainty. In *Recent developments and new directions in soft computing, volume 317 of Studies in fuzziness and soft computing*, ed. L.A. Zadeh, A.M. Abbasov, R.R. Yager, S.N. Shahbazova, and M.Z. Reformat, 393–407. Springer International Publishing.
28. Duda, J. 2005. Lot-sizing in a foundry using genetic algorithm and repair functions. In *Evolutionary computation in combinatorial optimization*, ed. G.R. Raidl, and J. Gottlieb, 101–111, Lecture Notes in Computer Science Berlin: Springer.
29. Michalewicz, Z., and C.Z. Janikow. 1991. Genetic algorithms for numerical optimization. *Statistics and Computing* 1(2): 75–91.

Chapter 7
Application of Fuzzy Decision Trees in Analog Forecasting

Abstract This chapter presents a new method for forecasting the level and structure of market demand for industrial goods. The method employs two data mining methods: k-means clustering and fuzzy decision tree learning. The k-means method serves to separate groups with items of a similar consumption level and structure of the analysed products. Whereas, fuzzy decision tree learning are used to determine the dependencies between consumption patterns and predictors. The proposed method is verified using the extensive statistical material on the level and structure of steel products consumption in selected countries in the years 1960–2010.

Analog forecasting is a method of forecasting a given variable by using information about the behaviour of another variable whose changes over time are similar, but not simultaneous. In this chapter a new method for forecasting the level and structure of demand based on the concept of analog forecasting is proposed. The method uses the concepts underlying historical and geographical analogies. It involves the estimation of the level and structure of demand based on the analysis of historical data and the use of data clustering and decision trees. The *k*-means method is used to separate groups of items having similar consumption level and similar structure of the analysed products (consumption patterns), whereas the fuzzy ID3 algorithm is used to build a fuzzy decision tree, i.e., to determine dependencies between consumption patterns and predictors (parameters determining the level and structure of consumption).

The proposed method can primarily be used to forecast the level and structure of demand for products that are traded on the industrial market. The proposed concept can, therefore, be used to forecast demand for products such as: steel industry products, non-ferrous metals industry products, certain chemical industry products, casts, construction materials industry products, energy carriers.

The investigations focus on the long-term forecasting method of demand for steel products, in which case forecasts for the level and structure of apparent consumption of steel products are produced. Apparent consumption is calculated as the difference between domestic production of the analysed range of steel products and the consumption of these products for further processing in the steel industry. This calculated difference is adjusted by the balance of foreign trade for these products [1].

© Springer International Publishing Switzerland 2015

I. Skalna et al., *Advances in Fuzzy Decision Making*,
Studies in Fuzziness and Soft Computing 333,
DOI 10.1007/978-3-319-26494-3_7

The presented concept, which has been verified on the example of steel products, can be applied, after modifications, to forecasting the demand for the abovementioned products, however, certain conditions must be met:

- historical data regarding the level and structure of consumption must be available,
- data regarding the value of predictors, which define the level and structure of demand for a given product, must be determined and available.

7.1 Methods for Forecasting Apparent Consumption of Steel Products

The following methods are used for forecasting apparent consumption of steel products:

- econometric models,
- sectoral analysis,
- trend estimation models,
- analog methods.

In econometric models, the level of apparent consumption of steel products is a function of selected macroeconomic parameters. Typically, gross domestic product (GDP), GDP composition, the value of investment outlays, and the level of industrial production are used as exogenous variables [1–6]. And often in econometric models GDP steel intensity is used as the endogenous variable. The term GDP steel intensity was defined by Malenbaum [7] as the amount of steel products consumed per one unit of GDP. The dependence of GDP steel intensity on selected macroeconomic parameters was used to forecast steel consumption in selected countries [1, 2]. Evans analysed GDP steel intensity of Britain's GDP in the postwar period. He accepted that GDP steel intensity depends on the share of industry and construction in making up the GDP, high technology industries of the GDP produced by the industry as a whole, the scope of use of material-saving technologies in industry and construction, and the substitution of steel with other materials.

Tilton [8] applied the sectoral analysis to forecast steel products consumption. It involves forecasting steel intensity indicators for selected sectors of the economy and the GDP generated in those sectors. Crompton [9] used the method proposed by Tilton for the analysis and forecast of steel consumption in Japan. Tilton's method was also used for forecasting steel consumption in the world [6], in the U.S. [5] and in Poland [1].

Apparent consumption of steel products was also forecasted by use of trend estimation models [10]. The forecasts produced by these methods are usually short-term forecasts.

Analog methods are also used for forecasting apparent consumption of steel products. They are based on the assumption that the indicators characterising the level and structure of the apparent consumption of steel products in the country for which the forecast is drawn up tend to attain the indicators characterising selected countries that serve as comparator. The indicators most often compared include consumption of steel products *per capita*, GDP steel intensity, and the assortment structure of apparent consumption [1].

7.2 Methodology

During the last decade, data mining methods have become very popular tools supporting decision-making processes. These methods can be used to search for patterns in large data sets (data clustering) or for assigning new objects to existing patterns (data classification). As a result, they can be used to characterise the dependencies between the level and structure of demand (consumption patterns) and selected parameters (numerical or qualitative), which in this case perform the role of predictors. Such dependencies, if existing, can be used to forecast the level and structure of demand.

There are various data mining methods that can be used for building classification models. Among statistical data mining methods, the linear discriminant function model has been widely used in solving many practical problems [11]. However, the effectiveness of this method deteriorates when the dependencies between forecasted values and exogenous variables are very complex and/or non-linear [11]. Situations of this kind are often encountered in practice. In such cases the machine learning methods are more appropriate. Examples include algorithms for generating decision trees, neural networks, Bayesian networks, and genetic algorithms [10]. Each of these methods leads to different knowledge representation.

Among the referred to methods for building classification models, neural networks and algorithms for building decision trees are most commonly used [12–14]. The strength of neural networks lies in their many possibilities for generalising information contained in the analysed data sets. Decision trees, however, surpass neural networks in regards to ease of interpretation of obtained results. Rules generated by decision trees that assign objects to different classes are easy to interpret even for users unfamiliar with the problems of data mining [15, 16]. This characteristic is very important, because decision-makers in the industry prefer tools that operate based on algorithms understandable to all participants in the decision-making process and provide easily interpretable information.

There are many methods for generating data clusters. In addition to traditional statistical methods, among which the most popular are hierarchical methods and methods for optimising the initial division of sets, *Kohonen's Self Organizing Maps (SOM)* and genetic algorithms are also used. Among the methods for optimising the initial division of sets the most popular is k-means method. The k-means is one of the simplest unsupervised learning algorithms that solve the well known clustering problem. The procedure follows a simple and easy way to classify a given data set through a certain number of clusters (assume k clusters) fixed a priori.

Taking of above under the attention, from the very large number of data mining methods available, two were used in the proposed forecasting method: the k-means method, used for data clustering, and the *Fuzzy Interactive Dichotomizer 3 (FID3)*, used to build decision trees (data classification rules). The k-means method creates consumption patterns, while the *FID3* algorithm builds a decision tree that assigns specific patterns of consumption to exogenous variables (predictors).

7.2.1 The k-means Method

The k-means method is one of the most popular methods for cluster analysis. Given a set of observations $(x_1, x_2, \ldots, x_n)$, where each observation is a d-dimensional real vector, k-means clustering aims to partition the those observations into k ($k \leqslant n$) sets $S = S_1, S_2, \ldots, S_k$. The k-means method can also be used to optimise an initial division of items. The most common k-means algorithm uses an iterative refinement technique:

1. Randomly, arbitrarily, or using a different criterion select k cluster centres.
2. Calculate the distance between each item and cluster centres.
3. Assign each item to the closest cluster.
4. Determine the centroids of the newly formed clusters.
5. If the stopping criterion is met Stop, otherwise Go To 3.

Subsequent iterations are characterised by the error function of items separation to individual clusters (*SES*) defined by the formula:

$$SES = \sum_{i=1}^{k} \sum_{j=1}^{n_i} d_{jS_i}^2, \tag{7.1}$$

where:

$d_{jS_i}^2$—the distance of the j-th item from the centroid of the i-th cluster,
n_i—the number of items belonging to i-th cluster.

Once the values of *SES* in subsequent iterations do not show significant changes (i.e., changes are less than the prescribed value) or when the maximum change of centroids (the distance of the new centroids from the previous ones) does not exceed the prescribed value, the procedure will stop (the stopping criterion is met). The stopping criterion can also be defined as the maximum number of iterations after which the process is stopped. The distance between centroids or between items is usually calculated using the Euclidean distance, which for x_i and y_j being two n-dimensional vectors, is defined as

$$d_E(x_i, y_j) = \sqrt{\sum_{k=1}^{n} \left(x_{ik} - y_{jk}\right)^2}. \tag{7.2}$$

Besides the Euclidean distance, the following two distance measures can also be used in k-means clustering:

- the root mean square distance (or average geometric distance):

$$d_{RMS}(x_i, y_j) = \sqrt{\frac{1}{n}\sum_{k=1}^{n}\left(x_{ik} - y_{jk}\right)^2}. \quad (7.3)$$

- Minkowski distance, which is a generalisation of Euclidean distance:

$$d_M(x_i, y_j) = q\sqrt{\sum_{k=1}^{p}\left(x_{ik} - v_{jk}\right)^q}, \quad (7.4)$$

where q is a positive integer. For $q = 1$ the Manhattan distance is obtained and for $q = 2$ the Euclidean distance.

7.2.2 Fuzzy Decision Trees

In the literature,several different algorithms of construction of crisp and fuzzy decision trees can be found. Fuzzy decision trees is the method, which combines fuzzy sets theory and fuzzy logic.

The most popular method for decision tree learning is ID3 algorithm which was initially proposed by Quinlan [17] and which was meant for making crisp decision trees. ID3 algorithm applies to a set of data and generates a decision tree for classifying the data. It uses the minimal entropy as a criterion at each node. The ID3 algorithm played a large role in the development of algorithms for building decision trees. The general idea of the tree induction algorithm is as follows:

- A tree starts with a single node representing the entire set of items.
- If all items belong to one decision class, the node becomes a leaf and is labelled with an appropriate decision.
- Otherwise the algorithm uses the information gain as a criterion for selecting the attribute that best divides the set of items.
- For the selected attribute a branch is created and the items are divided into the new nodes (subtrees).
- The algorithm further works recursively for the sets of items assigned to subtrees.
- The algorithm terminates when the stopping criterion is met.

The statistical quantity entropy is applied to define the information gain, to choose the best attribute from the candidate attributes. The definition of entropy is as follows:

$$H(S) = \sum_{i=1}^{N} -P_i \log_2 P_i \tag{7.5}$$

where P_i is the ratio of each clusters in set S (the number of elements in the cluster (subset) i divided by the number of all elements of the set S).

One of a disadvantage of decision tree is its instability. Decision tree is recognised as highly unstable classifier with respect to minor perturbations in the training data [18]. The structure of the decision tree may be entirely different if some things change in the data set. To overcome this problem, some scholars have suggested Fuzzy Decision Tree [19] by utilising the fuzzy set theory to describe the connected degree of attribute values, which can precisely distinguish the deference of subordinate relations between different examples and every attribute values [20].

The simple generalisation of ID3 algorithm on fuzzy numbers is so-called Fuzzy Interactive Dichotomizer 3 (FID3). The FID3 algorithm, is extended to apply to a fuzzy set of data (several data with membership grades) and generates a fuzzy decision tree using fuzzy sets defined by a user for all attributes. A fuzzy decision tree consists of nodes for testing attributes, edges for branching by test values of fuzzy sets defined by a user and leaves for deciding class names with certainties. Such an algorithm uses minimal fuzzy entropy or Information Gain as a criterion for decision-making. In the beginning, Fuzzy ID3 is only an extension of the ID3 algorithm achieved by applying fuzzy sets. It generates a fuzzy decision tree using fuzzy sets defined by a user for all attributes and utilises minimal fuzzy entropy to select expanded attributes. However, sometimes the result of this Fuzzy ID3 is poor in learning accuracy [19].

There are several variants of Fuzzy ID3 in the literature. For example one of them uses minimal classification ambiguity instead of minimal fuzzy entropy [21]. Other algorithms, which uses cumulative information estimations, were proposed by Levashenko and Zaiteva [22]. In these estimations we obtained Fuzzy Decision Trees with different properties (unordered, ordered, stable etc.).

Algorithms, which are quite different from ID3 were proposed for example by Wang et al. [23]. There are presented optimisation principles of fuzzy decision trees based on minimising the total number and average depth of leaves. Dong and Kothari [24] considered another non-Fuzzy ID3 algorithm where is used look-ahead method. Its goal is to evaluate so-called classifiability of instances that are split along branches of a given node.

Below is described the algorithm of FID3 used in the forecasting procedure. This algorithm is very similar to ID3, except ID3 selects the test attribute based on the information gain which is computed by the probability of ordinary data but this algorithm by the probability of membership values for data [19].

Assume that we have a set of data D, where each data has p numerical values for attributes $A_1, A_2, \ldots, A_p$ and one classified class $C = \{C_1, C_2, \ldots, C_n\}$ and fuzzy sets $\tilde{F}_{i1}, \tilde{F}_{i2}, \ldots, \tilde{F}_{il}$ for the attribute A_i (the value of l varies on every attribute). Let $\tilde{D}^{C_k}$ be a fuzzy subset in D whose class is C_k and let $|D|$ be the sum of the membership values in a fuzzy set of data D. Then an algorithm for generating a fuzzy decision

Algorithm 6 ID3 algorithm

1: Generate the root node that has a set of all data, i.e., a fuzzy set of all data with the membership value 1

2: If a node t with a fuzzy set of data $\tilde{D}$ satisfies the following conditions:

- the proportion of a data set of a class C_k is greater than or equal to a threshold Θ_r that is: $\frac{|\tilde{D}^{C_k}|}{|\tilde{D}|} \geqslant \Theta_r$,
- the number of a data set is less than a threshold Θ_n that is $|\tilde{D}| \leqslant \Theta_n$,
- there are no attributes for more classification,

then t is a leaf node and assigned by the class name (more detailed method is described below)

3: If a node t does not satisfy the above conditions, it is not a leaf and the test node is generated as follows:

3.1. For each A_i, $(i = 1, 2, \ldots, p)$, calculate the information gain $G(A_i, D)$ (to be described below) and select the test attribute $A_{\max}$ with the maximum gain.

3.2. Divide D into fuzzy subsets $\tilde{D}_1, \tilde{D}_2, \ldots, \tilde{D}_l$ according to $A_{\max}$, where the membership value of the data in $\tilde{D}_j$ is the product of the membership value in D and the value of $F_{\max,j}$ of the value of $A_{\max}$, in D.

3.3. Generate new nodes $t_l, t_2, \ldots, t_l$ for fuzzy subsets $\tilde{D}_1, \tilde{D}_2, \ldots, \tilde{D}_l$, and label the fuzzy sets $F_{\max,j}$ to edges that connect the nodes t_j and t.

3.4. Replace $\tilde{D}$ by $\tilde{D}_j$ $(j = 1, 2, \ldots, l)$ and repeat from recursively from 2.

tree is the following: The information gain $G(A_i, \tilde{D})$ for the attribute A_i by a fuzzy set of data D is defined by

$$G(A_i, \tilde{D}) = I(\tilde{D}) - E(A_i, \tilde{D}),$$

where:

$$I(\tilde{D}) = -\sum_{k=1}^{n} \frac{|\tilde{D}^{C_k}|}{D} \log_2 \frac{|\tilde{D}^{C_k}|}{\tilde{D}} \tag{7.6}$$

$$E(A_i, \tilde{D}) = \frac{|\tilde{D}_{F_{ij}}|}{\sum_{j=1}^{m} \tilde{D}_{F_{ij}}} I(\tilde{D}_{F_{ij}}) \tag{7.7}$$

As for assigning the class name to the leaf node, are proposed three methods as follows:

1. The node is assigned by the class name that has the greatest membership value, that is, other than the selected data are ignored.
2. If the condition (a) in step 2 in the algorithm holds. Do the same as the method (a). If not, the node is considered to be empty, that is, the data are ignored.
3. The node is assigned by all class names with their membership values, that is, all data are taken into account.

7.3 The Proposed Forecasting Method

The overview of forecasting methods conducted in Sect. 7.1 shows that the structure and level of apparent consumption of steel products or *GDP* steel intensity depend on selected macroeconomic parameters characterising a country's economy. The following were most often used as exogenous variables: the value of *GDP per capita* and the sectoral composition of the *GDP*. The sectoral composition is characterised by the contribution of industry and construction to *GDP*. Additionally, in the case of industry, the share of sectors determining the level of steel consumption (so-called steel intensity industries) in the industry is significant to the total *GDP*. Such sectors are: manufacture of finished metal products excluding machinery and equipment, manufacture of electrical equipment, manufacture of machinery and equipment not classified elsewhere, manufacture of motor vehicles and trailers, excluding motorcycles and the manufacture of other transport equipment.

The proposed method is used to forecast the following (endogenous variables in the model):

- GDP steel intensity (St),
- share of various ranges of products in consumption
  - share of long products (Ud),
  - share of flat products (Up),
  - share of pipes and hollow sections (Ur)
  - relation of the consumption of organic coated sheets to the consumption of metallurgic products altogether (Uo).

The exogenous variables in the model (predictors) are:

- value of GDP *per capita* (GDP),
- share of generated gross value in industry and construction in making up GDP (UPB),
- share of steel intensity industries in making up gross value of industry (UPS).

Endogenous variables create a consumption profile for each country in a particular year. Besides a consumption profile, the historical data include a summary of values of predictors (exogenous variables) for each country and year. The historical data covering consumption profiles and values of predictors are from different countries and different periods. For certain countries, only values of predictors were available in the forecasted period. Consumption profiles are forecasted on the basis of these values.

An example of historical data for a single country in one year is presented in Fig. 7.1.

The proposed forecasting procedure is as follows. First, consumption patterns are created. A consumption pattern is understood as *GDP* steel intensity and structure of consumption expressed by the share of individual ranges of steel products in total consumption. Such patterns are defined using data on consumption profiles of different countries over many years. The *k*-means method is used for this purpose. The

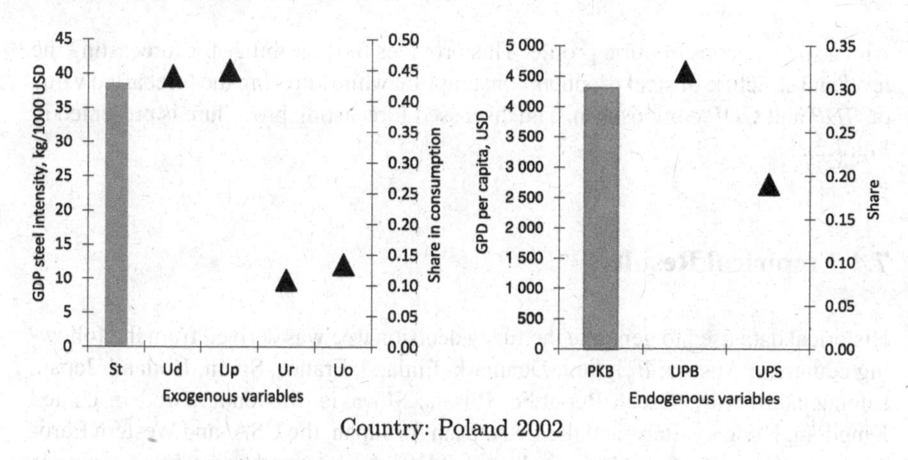

Fig. 7.1 An example of historical data used in the forecasting process

centroids of identified clusters make up the consumption patterns. After defining clusters and their centroids, a fuzzy decision tree is built based on which consumption patterns (the value of *GDP* steel intensity and consumption structure) can be assigned to certain values of the predictors (the value of *GDP per capita* and *GDP* composition).

In the proposed method historical data are used to build a fuzzy decision tree distinguishing easily interpretable links between predictors and pre-defined consumption patterns. The fuzzy decision tree allows for combining a vector of values of predictors

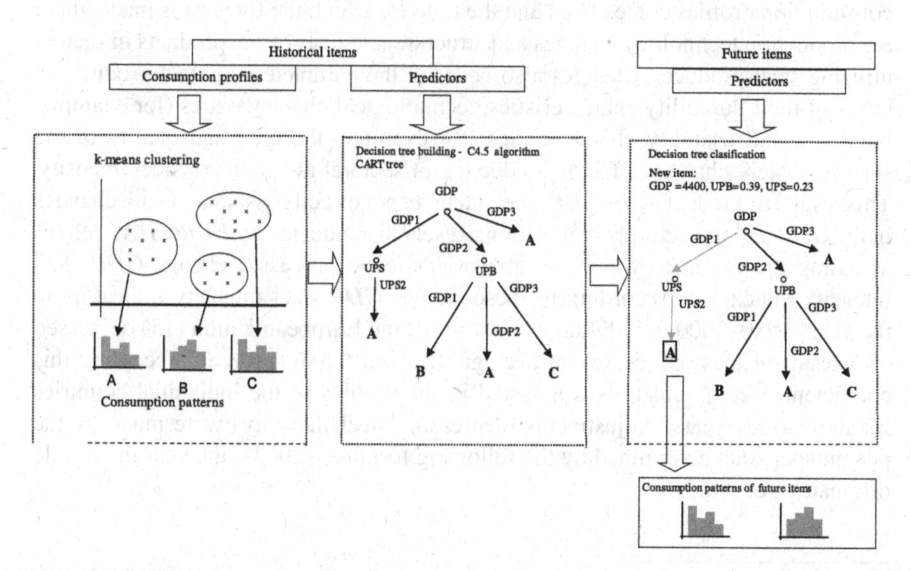

Fig. 7.2 The proposed forecasting procedure

with a specific consumption profile. This provides the possibility for forecasting the level and structure of steel products consumption while knowing the forecasted value of *GDP* and *GDP* composition. The discussed forecasting procedure is presented in Fig. 7.2.

7.4 Empirical Results

Historical data used to generate the fuzzy decision tree was derived from the following countries: Austria, Belgium, Denmark, Finland, France, Spain, Holland, Japan, Lithuania, Norway, Czech Republic, Russia, Slovakia, Slovenia, Sweden, United Kingdom, Hungary, Italy, and the USA. Data for Japan, the USA, and Western European countries came from the period 1960–2010, and data for the remaining countries came from the period 1993–2010. It was not possible to gather data on the structure of consumption for all counties in all years of the periods indicated above. Taking into account the gaps in the data for some of the years, there was a total of 730 items collected. The data was divided into two sets: a set of 667 items, which served to build a classifier and a set of 73 items, which were used to test the classifier and compare it with other method. The division of the set was conducted at random. Data on *GDP* for each country was expressed in dollars according to prices from 2007.[1]

The values of *GDP* steel intensity and demand structure defined by consumption patterns are used in determining the forecast in the proposed method. These patterns are defined on the basis of historical data describing the consumption profiles of various countries and periods. Between the time which the data characterising the consumption profiles comes from and the time for which the forecast is made, there are production technology changes and structural changes of the products in sectors utilising steel products. Changes also occur in the parameters of steel products in terms of their durability characteristics, technological characteristics (for example, bonding), and usability characteristics (for example, the type and quality of the surface). These changes affect the reduction of sectoral indicators of steel intensity. This results in a reduction of *GDP* steel intensity not directly associated with changes in the *GDP* and its sectoral composition. Research conducted by *EUROSTAT* allows assessing the extent to which the above-mentioned processes change *GDP* steel intensity indicators. According to these studies *GDP* steel intensity indicators in the years 1980–2000 in individual countries of the European Union (15) decreased as a result of these processes an average of about 0.5 % per year. Accepting this coefficient, steel intensity was adjusted in the profiles of the individual countries for the various years. Adjustments (decreasing steel intensity) were made by the percentage value determined by the following formula: (2007—the year the profile originates from) 0.5 %.

[1] The sources of data were corresponding yearly publications: The Steel Market, Annual International Statistics, and annual statistics of individual countries.

In the analysed example, the share of gross value produced by steel intensity industries as part of the gross value produced by the industry in general was accepted as one of the exogenous variables. This assumption is justified when the profiles take into account the structure of consumption by major groups of ranges of steel products (long products, flat products, pipes and organic coated sheet). Deeper disaggregation of consumption is likely to require determining *GDP* composition divided by individual steel intensity industries. This is the subject of further research. The main problem is obtaining reliable data on the consumption of particular ranges of steel products in different countries.

In the first step one performed the clustering of steel intensity and demand structure for the purpose of the obtainment of patterns of the consumption. In the process of the clustering one accepted 9 classes. Results of the clustering are presented in the Table 7.1 (values after the normalisation).

Table 7.2 presents the characteristics of individual clusters.

To build FID3, the fuzzification of all independent variables was carried out. Fuzzification relied on the division of every attribute into classes of the same size. The affiliation to the classes is described by fuzzy number. To the fuzzification one used the following algorithm. The attribute A is a sequence $J+1$ observation $A = \{x_j\}_{j=0,1,\ldots,J}$ such that $x_{j+1} \geqslant x_j$. Observations are divided into l disjoint

Table 7.1 Result of the clustering of structures of the consumption

	St	Ud	Up	Ur	Uo
1	0.184083	0.343596	0.651336	0.431067	0.507877
2	0.53204	0.442453	0.546201	0.502227	0.226074
3	0.39037	0.53588	0.550116	0.152016	0.314489
4	0.737196	0.514734	0.438294	0.665591	0.177773
5	0.31663	0.567211	0.46416	0.38451	0.072881
6	0.334618	0.549177	0.498734	0.311504	0.494481
7	0.313672	0.673825	0.360576	0.420971	0.33778
8	0.113727	0.268793	0.752175	0.303377	0.68626
9	0.529184	0.788603	0.261072	0.407442	0.393627

Table 7.2 The characteristics of individual classes

Cluster	1	2	3	4	5	6	7	8	9
Number of items	118	61	73	56	85	105	76	90	66
Average									
St, kg/1000USD	15.702	35.826	27.632	47.691	23.368	24.408	23.197	11.633	35.66
Ud	0.396	0.418	0.439	0.435	0.447	0.442	0.471	0.379	0.497
Up	0.504	0.477	0.478	0.449	0.456	0.465	0.429	0.53	0.404
Ur	0.1	0.105	0.083	0.115	0.097	0.093	0.1	0.092	0.099
Uo	0.2	0.129	0.152	0.117	0.091	0.196	0.157	0.244	0.171

subsets ($S_k, k = 1, 2, \ldots, l$), where $l = 5 + 3.3 * \log(J) + 1$ (in the example presented here $l = 11$). Then, a trapezoidal membership function $\tilde{H}_k = (a_k, b_k, c_k, d_k)$ is determined for each subset S_k, where values a_k, b_k, c_k, d_k are designated in the following way:

$k = 1$	$\tilde{H}_k = \begin{cases} a_k = g_k \\ b_k = g_k \\ c_k = g_{k+1} \\ d_k = g_{k+1}(1+m) \end{cases}$
$1 < k < l+1$	$\tilde{H}_k = \begin{cases} a_k = g_k(1-m) \\ b_k = g_k \\ c_k = g_{k+1} \\ d_k = g_{k+1}(1+m) \end{cases}$
$k = l+1$	$\tilde{H}_k = \begin{cases} a_k = g_k(1-m) \\ b_k = g_k \\ c_k = g_{k+1} \\ d_k = g_{k+1} \end{cases}$

where $g_k = x_{j_k}$, $j_k = (k-1)\left\lfloor \frac{J+1}{l} \right\rfloor$, $k = 1, \ldots, l+1$ and $m \in [0, 1]$.

The following values of m were considered:

	m (%)
GDP	5
UPB	1
UPS	4

Further, specific class names were used. Class names for variables are: the class name(i), where i is number of the class, e.g., *GDP1* (the first class for variable *GDP*).

Then, the decision tree of the type *CART* was built using the algorithm *FID3* described in Sect. 7.2.2.

In the process of construction of a fuzzy decision tree, the following stopping criterion was employed:

	Threshold (%)
Θ_r	3.00
Θ_n	90.00

The final decision tree consists of 216 leaves and 100 nodes. The *GDP* attribute is the most rarely tested. This means that it is the least important in determining the level and structure of steel products consumption. Sectoral composition of the *GDP* is more relevant in this case.

Table 7.3 The example-case for prediction

GDP		BUD	STAL	
GDP7	*GDP8*	*BUD2*	*STAL7*	*STAL8*
0.22	0.78	1	0.54	0.46

Inference in a ordinary decision tree is executed by starting from the root node and repeating to test the attribute at the node and branch to an edge by its value until reaching at a leaf node, a class attached to the leaf being as the result. The difference between the ordinary and fuzzy tree relies on this that the given case is not credited only to one branch, but to many with some degree of the membership.

For example, Table 7.3 presents the example-case. On the basis this case will be elaborated the forecast. For simplicity, only non-zero values of the membership function are shown below.

To elaborate the forecast the following three operations must be executed. First, for every branch one ought to designate the indicator *F*—this is the value of membership function with which the attribute of the case for which we elaborate the forecast satisfies the rule by which the branch is based. Figure 7.3 presents a part of the obtained fuzzy decision tree, where values of indicators *F* for individual branches are non-zero. Values of indicators *F* are placed in the grey rectangles near the respective branches. In the second step is designated membership function for analysed case to individual clusters. Below one showed suitable calculations.

In the third step on the basis of such a defined membership function for the analysed case is designated forecast.

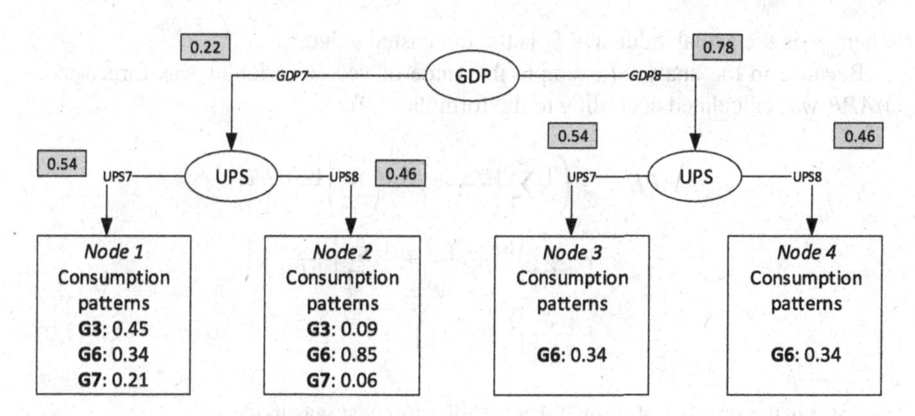

Fig. 7.3 Part of the obtained fuzzy decision tree

$$0.220.54\begin{vmatrix}0.45\\0.35\\0.20\end{vmatrix}+0.220.46\begin{vmatrix}0.09\\0.85\\0.06\end{vmatrix}+0.780.54\begin{vmatrix}0.00\\0.34\\0.00\end{vmatrix}+0.780.54\begin{vmatrix}0.00\\0.34\\0.00\end{vmatrix}=\begin{vmatrix}0.06\\0.90\\0.04\end{vmatrix}\begin{matrix}G3\\G6\\G7\end{matrix}$$

	G3	G6	G7	Forecast
St, kg/1000USD	27.632	24.408	23.197	**24.553**
Ud	0.439	0.442	0.471	**0.443**
Up	0.478	0.465	0.429	**0.464**
Ur	0.083	0.093	0.1	**0.093**
Uo	0.152	0.196	0.157	**0.192**

7.5 The Evaluation of Obtained Results

The quality of the classifier built in the form of a fuzzy decision tree can be assessed by various indicators. In the analysed example, the values of continuous variables are forecasted. As a measure of accuracy, *the mean absolute percentage error (MAPE)* is used, defined by the formula:

$$MAPE = \frac{1}{n}\sum_{t=1}^{n}\left|\frac{y_i-\hat{y}_i}{y_i}\right| \tag{7.8}$$

where y_i is the actual value and $\hat{y}_i$ is the forecasted value.

Because in the analysed example the value of vector variables was forecasted, *MAPE* was calculated according to the formula (7.9).

$$\begin{aligned}MAPE = \frac{1}{5}\Bigg(&\frac{1}{n}\sum_{i=1}^{n}\left|\frac{St_i-\hat{S}t_i}{St_i}\right| + \frac{1}{n}\sum_{i=1}^{n}\left|\frac{Ud_i-\hat{U}d_i}{Ud_i}\right|\\ &+\frac{1}{n}\sum_{i=1}^{n}\left|\frac{Up_i-\hat{U}p_i}{Up_i}\right| + \frac{1}{n}\sum_{i=1}^{n}\left|\frac{Ur_i-\hat{U}r_i}{Ur_i}\right|\\ &+\frac{1}{n}\sum_{i=1}^{n}\left|\frac{Uo_i-\hat{U}o_i}{Uo_i}\right|\Bigg)\end{aligned} \tag{7.9}$$

where n is the number of items for which the forecast was made.

It is difficult to compare the proposed method with conventional econometric models, since a vector characterising the level and structure of consumption is forecasted. The vector components must meet certain conditions. The sum of the forecasted shares of the three main product groups (long products, flat products, pipes) must be 1. In order to compare the effectiveness of the proposed fuzzy method an additional classifiers were also tested based on traditional (crisp) *C&RT* method which used the *Gini* coefficient to build the tree.

Table 7.4 Comparison of the *mean absolute percentage error* between the 2 tested models

	Mean absolute percentage error
The proposed method (fuzzy decision tree)	0.028
The CART crisp method	0.076

A comparison of the quality of the forecasts was made using 67 selected items. The quality of the forecasts was rated by calculating *MAPE* according to the formula (7.9). The results of the tests are presented in Table 7.4.

The data in Table 7.4 indicate that the proposed fuzzy algorithm provides the more accurate forecasts of the level and structure of consumption. This algorithm provided the smaller value of the *mean absolute percentage error*. In this case, the value of the *mean absolute percentage error* constitutes 36.8 % values of the error in the crisp *CART* method using the *Gini* coefficient to select attributes based on which the training set is divided.

The presented concept provides good results, which allow it to be recommended as one of the methods for long-term forecasting of demand for selected products traded on the industrial market. This method can be classified into the group of analog methods. In such methods a forecast is formulated on the basis of comparator. Most comparators are defined by experts. The proposed method allows for objectifying the selection of comparators by making the selection dependent on the values of chosen predictors. The method makes it possible to forecast the level and structure of demand.

References

1. Rębiasz, B. 2006. Polish steel consumption 1974–2008. *Resources Policy* 31(1): 37–49.
2. Evans, M. 1996. Modeling steel demand in the UK. *Ironmaking and Steelmaking* 23(1): 19–23.
3. Ghosh, S. 2006. Steel consumption and economic growth: Evidence from India. *Resources Policy* 31(1): 7–11.
4. Labson, B.S. 1997. Changing patterns of trade in the world iron ore and steel market: An econometric analysis. *Journal of Policy Modeling* 19(3): 237–251.
5. Roberts, M.C. 1990. Predicting metal consumption: The case of US steel. *Resources Policy* 16(1): 56–73.
6. Roberts, M.C. 1996. Metal use and the world economy. *Resources Policy* 22(3): 183–196.
7. Malenbaum, W. 1975. *World demand for raw materials in 1985 and 2000*. New York: McGrow-Hill.
8. Tilton, J.E. 1986. Beyond intensity of use. *Materials and Society* 10(3): 1–14.
9. Crompton, P. 2000. Future trends in Japanese steel consumption. *Resources Policy* 26(2): 1–14.
10. Thomassey, S., and A. Fiordaliso. 2006. A hybrid forecasting system based on clustering and decision trees. *Decision Support Systems* 42(1): 408–421.
11. Altman, E.I., G. Macro, and F. Varetto. 1994. Corporate distress diagnosis: Comparison using linear discriminant analysis and neural networks (the Italian experience). *Journal Banking and Finance* 18(3): 505–529.

12. Andrès, J., M. Landajo, and P. Lorca. 2005. Forecasting business profitability by using classification technique: A comparative analysis based on Spanish case. *European Journal of Operational Research* 167(2): 518–542.
13. Rastogi, R., and K. Shim. 2000. Public: A decision tree classifier that integrates building and pruning. *Data Mining and Knowledge Discovery* 4(4): 315–344.
14. Tsujimo, K., and S. Nishida. 1995. Implementation and refinement of decision tree using nneural network for hybrid knowledge acquisition. *Artificial Intelligence in Engineering* 9(4): 265–275. Selected papers from the 1994 Japan/Korea joint conference on expert systems.
15. Hansen, L.K., and P. Salamon. 1990. Neural network ensembles. *IEEE Transaction on Pattern Analysis and Machine Intelligence* 12(10): 993–1001.
16. Zhou, Z.-H., and Y. Jiang. 2004. Nec4.5: Neural ensemble based c4.5. *IEEE Transaction on Knowledge and Data Engineering* 16(6): 770–773.
17. Quinlan, J.R. 1986. Induction of decision trees. *Machine Learning* 1(1): 81–106.
18. Olaru, C., and L. Wehenkel. 2003. A complete fuzzy decision tree technique. *Fuzzy Sets and Systems* 138(2): 221–254.
19. Umanol, M., H. Okamoto, I. Hatono, H. Tamura, F. Kawachi, S. Umedzu, and J. Kinoshita. 1994. Fuzzy decision trees by fuzzy ID3 algorithm and its application to diagnosis systems. In *proceedings of the third IEEE conference on Fuzzy systems, IEEE world congress on computational intelligence*, vol. 3, 2113–2118, Orlando, FL, June 26–29. IEEE World Congress on Computational Intelligence.
20. Chandra, B., and P.P. Varghese. 2007. On improving efficiency of SLIQ decision tree algorithm. In *International joint conference on neural networks, IJCNN 2007*, 66–71.
21. Yuan, Y., and M.J. Shaw. 1995. Induction of fuzzy decision trees. *Fuzzy Sets and Systems* 69(2): 125–139.
22. Zaitseva, E., V. Levashenko, K. Pancerz, and J. Gomuła. 2014. Fuzzy decision tree based classification of psychometric data. In *Position papers of the 2014 federated conference on computer science and information systems, ACSIS*, vol. 3, 37–41.
23. Wei, J.-M., M.-Y. Wang, J.-P. You, S.-Q. Wang, and D.-Y. Liu. 2007. VPRSM based decision tree classifier. *Computing and Informatics* 26(6): 663–677.
24. Dong, M., and R. Kothari. 2001. Look-ahead based fuzzy decision tree induction. *IEEE Transactions on Fuzzy System* 9(3): 461–468.

Chapter 8
Selected Issues of Visualisation of Fuzziness in Cardiac Imaging Data

Abstract This chapter proposes an approach to visualisation of fuzzy numbers in one-, two- and more-dimensional spaces. The proposed approach is based on ScPovPlot3D templates for POVRay.

In the recent years technological development enabled elaboration of numerous outstanding data gathering methods [1] which produces large amounts of data in both numeric and raster image format. Amongst most productive disciplines one can mention astronomy, engineering, economics, econometrics, automatics, medical sciences including medical imaging. Reliable analysis of obtained data more often than not requires the usage of computer processing and then data visualisation. Accordingly to the subject of this book, processed data are assumed to be fuzzy, so special techniques of visualisation are required for both input and processed (output) data.

8.1 Visualisation of Fuzzy Numbers

Chapter 1 presents several pictures illustrating main concepts of interval and fuzzy numbers. Such form, while being communicative, is highly impractical regarding visualisation of non-trivial sets of fuzzy numbers, such as one and multi-dimensional series of measurements, solutions of differential equations with fuzzy parameters or fuzzy boundary conditions or computed or measured fuzzy surfaces [2].

8.1.1 Visualisation of a Fuzzy Number

One dimensional fuzzy point is simply a fuzzy number. It can be visualised in the well-known form of interval (Fig. 8.1) with whiskers and if membership function is to be presented, as a Box and Whiskers symbol (Fig. 8.2). However, according to the interval or fuzzy value paradigm no uniquely defined "central" or "most important"

© Springer International Publishing Switzerland 2015

I. Skalna et al., *Advances in Fuzzy Decision Making*,
Studies in Fuzziness and Soft Computing 333,
DOI 10.1007/978-3-319-26494-3_8

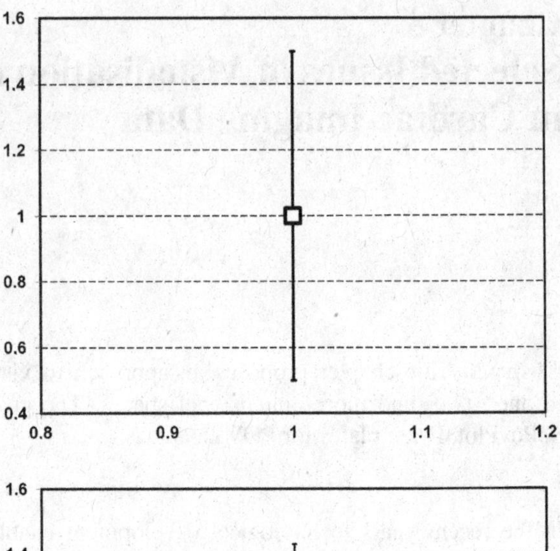

Fig. 8.1 Traditional visualisation of one dimensional fuzziness of data. In this case, 'x' coordinate is known exactly (it may be just a measurement number) but 'y' coordinate is determined with limited accuracy. Usually, central point, represented here by symbol of *square*, points at hypothetical accurate 'y' value while whiskers shows level of uncertainty. *Source* own

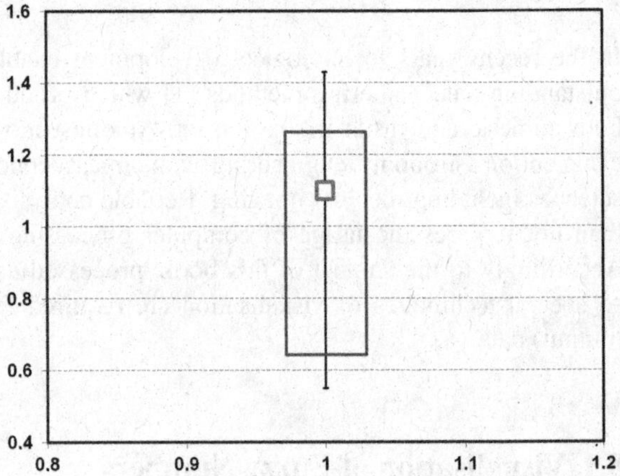

Fig. 8.2 Traditional visualisation of one dimensional fuzziness of data including internal distribution or membership function. The *box* represents for example range of standard deviation while *whiskers* shows *lower* and *upper* boundary of the fuzzy number

point exists. As stated in the first and second chapters, this is the main reason why relational operators cannot be uniquely defined.

Fuzzy symbols shown in the both pictures may be modified in several ways. For example one may add additional whiskers showing end points of consecutive α-cuts of triangular fuzzy number (see Sect. 1.1, Definition 1.5, Fig. 1.5) or core and support values of trapezoidal fuzzy number (see Sect. 1.1, Definition 1.6, Fig. 1.7). Thus shape and properties of membership function can be effectively visualised, even on scattered plots composed of several fuzzy values.

In this context Japanese Candlesticks, commonly used for technical analysis of trends on stock markets (see [3]), may be assumed as visualisation technique for time dependent fuzzy numbers (Fig. 8.3).

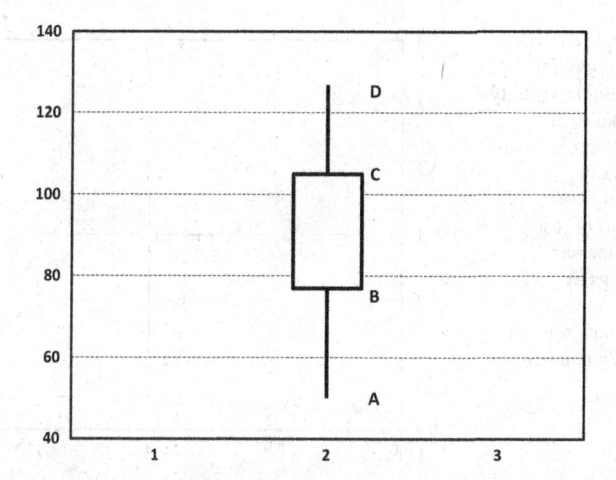

Fig. 8.3 Example of Japanese Candlestick chart (one session). In the picture *white candlestick* is shown. In this case *central box* illustrates opening (B) and closing (C) prices while ends of *vertical line* segments depicts session maximum (D) and minimum (A), however price dynamic during session cannot be shown in this manner. Several candlesticks may create specific formations which, if spotted, might be helpful for share traders at buy-keep-sell decision making. *Source* own

8.1.2 Visualisation of a Two Dimensional Fuzzy Point

Visualisation of a two-dimensional fuzzy point may be done by simple extension of one dimensional case. As shown in Fig. 8.4, both, vertical and horizontal uncertainty may be shown together by means of "Box and Whiskers" style applied to both dimensions respectively. The size of the internal rectangle is usually connected with the standard deviation, but in this case it may be interpreted as the "core" of a fuzzy number, whereas the whiskers may depict the "support" of a fuzzy number. As in the one-dimensional case, marking of "central" makes no sense.

The described solution seems to be obvious and quite descriptive. However there is a serious flow in such approach. Box and whiskers mark exact values of kernel and support around "central lines" only. Support and kernel values measured along intermediate directions are expected to be located on some kind of curve rather than on the straight line. Thus, by analogy to multidimensional error analysis, ellipses, as in Fig. 8.5 are proposed as approximate boundaries, i.e., support and kernel curves,[1] respectively. More detailed analysis may suggest other curves family in the future. In this example fuzzy ellipse is aligned to XY axes, but in general other orientations cannot be excluded á priori.

Usually, measurements or calculations produce series of data, which in this book are assumed to be fuzzy points. In order to visualise relations between them scatter

[1]In statistics *error ellipse* is connected with covariance matrix of two stochastic, potentially correlated variables. While this problem is out of scope of this book, it is thoroughly discussed in relevant literature. Excellent explanation is given on the page [4].

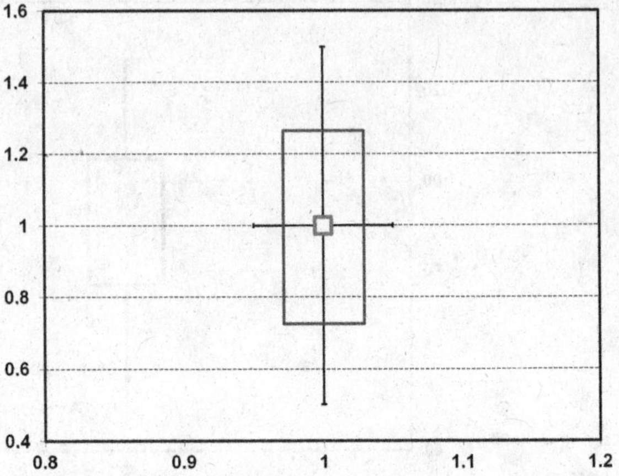

Fig. 8.4 Example of two dimensional fuzzy point. Small *central square* visually pinpoints symbol to the chart. Larger *rectangle* gives clue on uncertainty distribution along *horizontal* and *vertical axes* or just shape of a membership function. Endings of *horizontal* and *vertical line* segment may show support of relevant fuzzy number

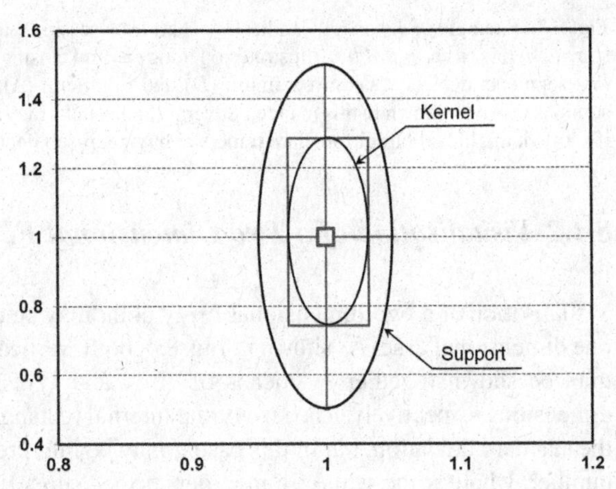

Fig. 8.5 Example of two dimensional fuzzy point. Small *central square* visually pinpoints symbol to the chart. *Ellipses* marked as "Support" and "Kernel" represents support and core boundaries of trapezoidal fuzzy point respectively

data plots may be employed. However there are no software packages supporting explicitly drawing charts with elliptic two-dimensional fuzzy points, but this may be easily changed in the future. Some of the packages may only roughly simulate this behaviour using bitmap symbols. In the latter case, unfortunately, all symbols ought to have the same size, what may not be relevant in every case (Fig. 8.6).

8.1.3 Visualization of a Three-Dimensional Fuzzy Points

In three dimensions Box and Whiskers style is no longer suitable as there is no simple extension of line-art into 3D drawing. However ellipse style can be easily extended into more dimensional spaces as in 3-dimensional space it just takes shape of 3 axis ellipsoid (Fig. 8.7). Assuming that all components of the point in three

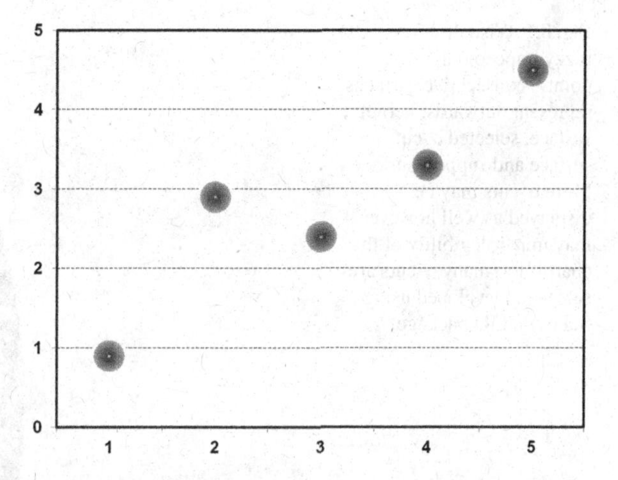

Fig. 8.6 Simulated fuzzy scattered data plot, employing symbols filled with radial gradient (Excel 2007)

dimensional space may be fuzzy numbers, real 3D chart is needed to visualise results of measurements or calculations. In such case, point data may be imaged as scattered fuzzy 3D chart, and surface data as fuzzy surface plot.

As mentioned above a dedicated software is necessary in order to ease drawing of fuzzy 3D plots. One of them[2] is a developer version of ScPovPlot3D package[3] built on top of POVRay,[4] well known ray-tracing program. ScPovPlot3D employs a domain specific language of POVRay called SDL (*Scene Description Language*), which makes it flexible and extensible solution. While thorough discussion of the package is out of the scope of this book, some examples are shown below.

8.2 Scattered Fuzzy 3D Chart

One of common cases where 3D fuzzy points may be found is a measurement series of three variables. Such series consist of ordered (for example by readout time[5]) or unordered list of 3D points indexed by natural numbers (Fig. 8.8).

In the presented example a short series has been computed using simple linear, but stochastically perturbed formula $P_i(x, y, z) = [2 + \delta; y_i, ay_i + b + \varepsilon]$ (see Eq. (8.1))

$$P_i(x, y, z) = \begin{cases} 2 + \delta, \\ y_i, \\ ay_i + b + \varepsilon . \end{cases} \tag{8.1}$$

---

[2]Authors have no information on availability of other packages.

[3]www.scpovplot3d.sf.net, accessed 2014-09-30.

[4]www.povray.org, accessed 2014-09-30.

[5]Such a case may be considered 4D example with time as an independent variable.

Fig. 8.7 Visualisation of 3D fuzzy trapezoidal point—consecutive surfaces represent outwards: kernel surface, selected α-cut surface and support surface. More α-cuts may be displayed as well however it may impair legibility of the chart if too many α-cuts are selected. Developed using ScPovPlot3D package

where δ and ε are random perturbations, a and b are coefficients, y_i is an independent variable of a measurement series and $i = 1, \ldots, N$. For testing purposes it has been assumed, that every measurement is in the form of 3D trapezoidal fuzzy number described above and length of semiaxis is equal to half of support value for all components. For x, y and z axis values 0.6, 0.2 and 0.8 respectively were adopted. Resulting picture is shown in Fig. 8.8.

Fuzziness of given data point may result from inaccuracy embedded into laboratory setup, assessment method, especially in social sciences, manufacturing inaccuracy (engineering, metallurgy) and sometimes from Heisenberg indeterminacy (uncertainty) principle (quantum phenomena).

Apart from visualisation of fuzzy numbers this technique may be applied to visualisation of dispersion of rigid numbers obtained in the course of repetitive measurements of single data point, for example diameter of a cylinder. In this case, consecutive layers of 3D fuzzy number may represent for example percentiles or multiples of standard deviation range. Furthermore, in case of correlation between measured x, y, z variables, a point-ellipsoid may be adequately rotated. In the latter case a covariance error ellipse procedure extended into three dimensions might be useful as well as solutions based on Mahalanobis distance (refer to [5, 6]).

In order to enhance visual message of the picture presented in the Fig. 8.8, right projections of fuzzy points may be added on base planes of coordinate system.

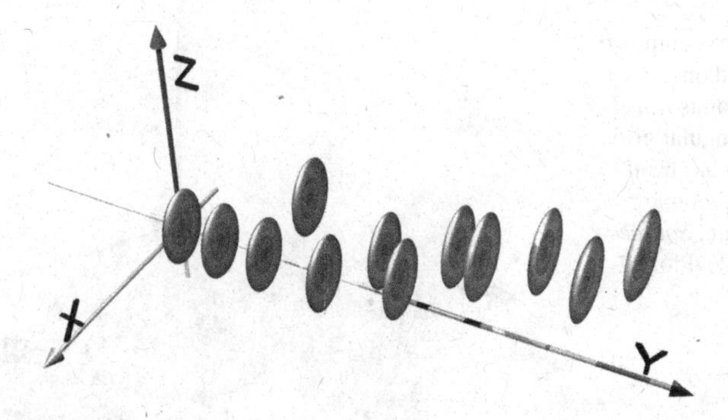

Fig. 8.8 Visualization of 3D fuzzy trapezoidal points—consecutive surfaces represent outwards: core surface, selected α-cut surface and support surface. Lengths of semiaxes of point-ellipsoids are proportional to their fuzziness: 0.6, 0.2, 0.8 measured along x, y and z axis respectively. Points are distributed due to the stochastically perturbated formula: $P_i(x, y, z) = [2 + \delta;\ y_i, ay_i + \mathrm{b} + \varepsilon]$. Developed using ScPovPlot3D package

8.2.1 Fuzzy Surface Plot

Problem of fuzzy surfaces visualisation, has been primarily addressed in the paper [2]. In the range of problems uncertainty is distributed over surface $z = \mathrm{f}(x, y)$,[6] usually defined over rectangular domain, but in fact, there is no such restriction in general. One of the sciences widely employing surface visualisation is geology which uses it to present distribution of mineral resources, principally ore deposits, for example coal or oil. Such a distribution may be probed using wide range of exploratory geophysical techniques like borehole logging, electrical resistivity tomography or gravimetry measurements. Unfortunately usually such mapping could not be done over rectangular grid required to display surface visualisation thus measurement points are freely distributed over the survey field. By applying numerical technique known as kriging (refer to [2, 7]) a regular grid may be evaluated, but values computed in every node are subject to uncertainty, resulting both from uncertainty of measured values and numerical, for example round-off, errors. As stated before, such uncertainty may be expressed in numerous ways, including standard deviation measure and of course fuzzy number paradigm. In the latter case first solution is to draw "thick surface", in contrast to normal, "infinitely thin surface" commonly used. Package ScPovPlot3D cannot explicitly implement "thick" or "fat" surfaces but this quality can be simulated utilising two features of POVRay: *prism* object and *media*[7] statement.

Media statement introduces fog resembling interior of an object with controlled density. Thick facets may be rendered with or without media statement, and accordingly, with opaque or transparent surface, fully or partially, so there is a lot

[6]Or more generally f(x, y, z) = 0.

[7]*Media* is a specific to POVRay implementation of *particles*.

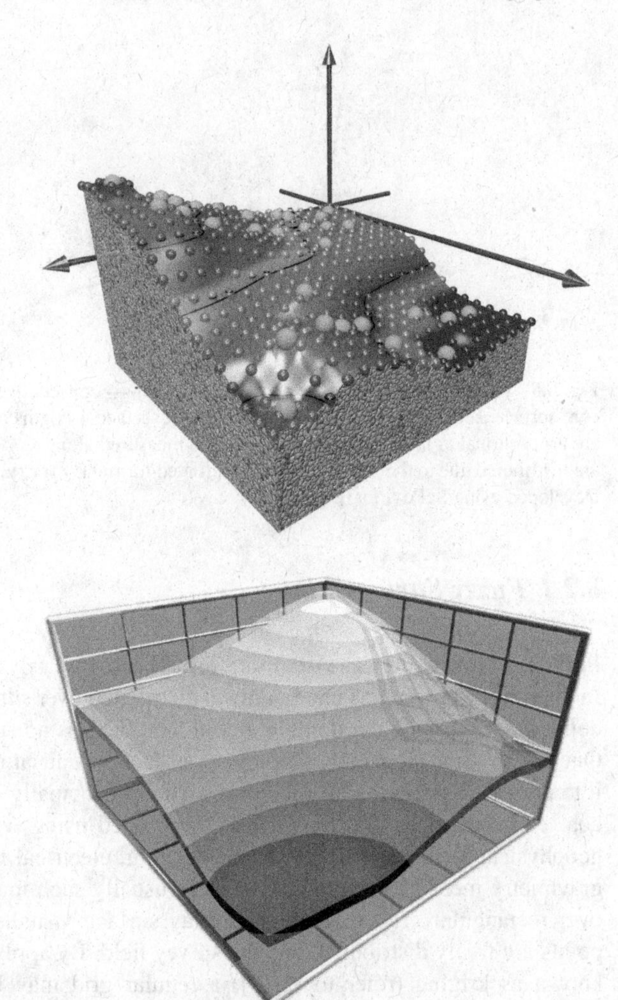

Fig. 8.9 Surface computed by kriging based on measurement points (*larger spheres*). Rectangular grid (*smallers balls*) has been computed using simple kriging algorithm. *Source* [2], own by ScPovPlot3D

Fig. 8.10 Example of fuzzy surface. Support range is represented by two separate *surfaces*, *lower* end displayed as *grayscale* mapped surface and the *upper* as unicolour but semitransparent one. Unfortunately implicit functions can not be represented this way. *Source* [2], visualisation created in ScPovPlot3D

of different arrangements and one of them may be suitable for a specific situation (Figs. 8.9 and 8.10).

The common practise in 3D computer graphics is splitting objects into meshes composed of myriads of tiny triangles, named *facets*. Specially crafted shadow functions are responsible for experience of smooth tonal transitions, as well as, simulating "smooth" look of the surface. By replacing every facet by prism with base aligned with the facet one can obtain "thick" surface elements comprising whole thick surface object. Of course information on the thickness of every element ought to be delivered for every facet separately. Some variants of thick facets are shown in Fig. 8.11 with relevant explanation.

In the next two sections another two examples of application of fuzzy surfaces will be presented. First one shows surface defined by fuzzy relationship $\tilde{z} = \tilde{z}(x, y)$,

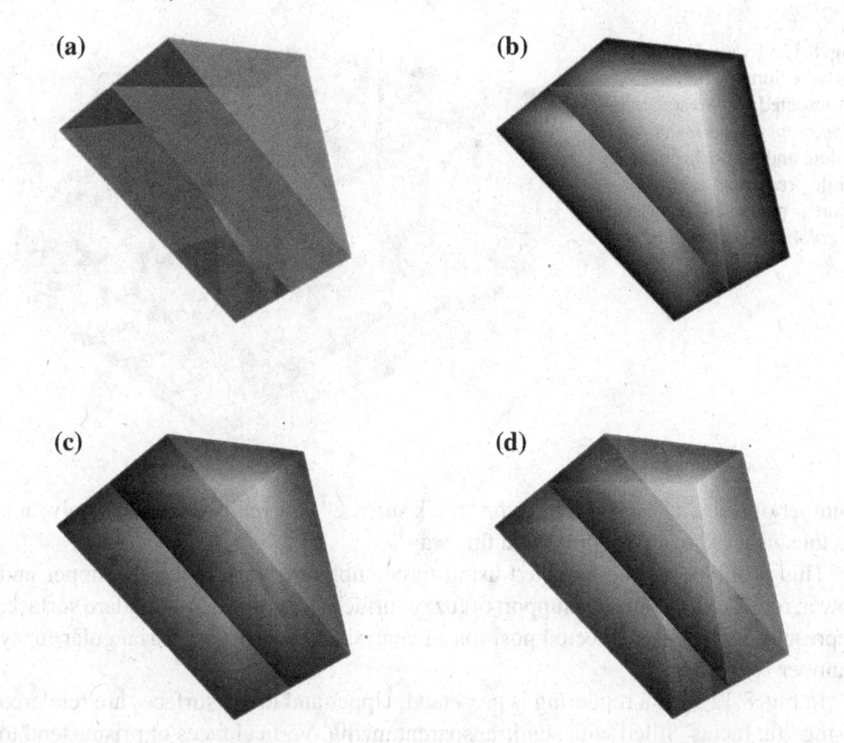

Fig. 8.11 Examples of fuzzy surface facets. Different settings has been applied to the same geometry. **a** Texture only and *central triangle*, **b** media only and *central triangle*, **c** media and texture without *central triangle*, **d** all components. Support range is represented by distance between triangular bases of the prism. Image produced using ScPovPlot3D package

where $\tilde{z}$ denotes fuzzy number. The second example shows lesions deteriorating blood flow deposited in a segment of cardiac vessel.

Example 1: Visualisation of $\tilde{z} = \tilde{z}(x, y)$ Surface

Implementation of thick surfaces using media (or *particles*) requires generation of complex polymeshes. This can be fully done using applications employing mesh as a generic object. Some of such applications can be listed: Blender, AutoCAD or 3DStudio. In every case efficient automated evaluation of the mesh requires sophisticated coding.[8] However, resulting object will render perfectly smooth.

In order to illustrate "thick surface" drawing style, the POVRay, popular raytracing program has been used. In this case, thick surface consists of a set of individual prism objects. As there is no obvious method of deformation of the prism to follow curvature of the surface (outer base should be wider than inner, and rarely symmetrically),

[8]For Blender efficient user space API written in Python is available.

Fig. 8.12 Example of fuzzy surface. Support range is represented by two separate *planes upper* and *lower* (dilute and monochromatic), while predicted "central" plane is presented in between as *color mapped surface*

number of artifacts is produced if the "thick surface" gets really thick, thus only "not so fat surfaces" can be represented this way.

This problem can be resolved using three supplementary surfaces—upper and lower, representing range of support of fuzzy surface and the middle, standard surface, representing visually expected position of central (modal) point, if triangular fuzzy number is employed.

In Fig. 8.12 such a rendering is presented. Upper and lower surfaces are rendered using "fat facets" filled with semitransparent media. Vertical faces of prisms tend to form artifacts and due to curvature of the surface, some cracks can be observed as well. However, despite these defects, the picture clearly represents fuzzy nature of related problem while artifact, as side effect, increase its readability.

In the end another problem should be addressed. Usage of media interior, rapidly increases compute time of the scene. Thus powerful multicore computer is required, possibly employing CUDA technology, happily a range of plugins to popular 3D modellers is available.[9]

8.2.2 Imaging of Fuzziness of Cardiac Vessels Walls

Angiography is a diagnostic method suitable for visualisation of geometry of coronary blood vessels and assessment of arteries walls. The method requires intravenous administration of radiocontrast agents and subsequent registration of X-Ray images in a classical or Computed Tomography (CT) setup [1, 9, 10]. The variant of angiography is a coronary angiography (CAG). In this case, the image of a vessel is obtained using catheter introduced into coronary vessel, which enables the localisation and assessment of extent of lesions and severity of coronary stenoses [11].

[9]For example, FurryBall GPU plugin renderer is available for Autodesk Maya, 3DS Max and Cinema 4D [8], nice comparison to another plugins is available at http://furryball.aaa-studio.eu/aboutFurryBall/compare.html.

This diagnostic method is invasive and dangerous for a patient. Therefore it should be performed only if really necessary.

Raw angiography does not allow for collecting detailed information on extent and severity of development of arteriosclerotic vascular disease, for example in the vicinity of the wall of a blood vessel [1, 10]. However, it is usually used for rough assessment of stenosis of coronary vessels, changes in walls of coronary arteries and anomalies in their placement, as well as, orifices [9]. Angiography, as opposed to coronary angiography (CAG) [12], is assumed to be safe, even for high risk patients [13]. However the risk connected with intravascular administration of radiocontrast agent still remains in patients with kidney disorder or failure.

Coronary angiography enhanced by computer analysis of image of coronary arteries (QCA, *Quantitative Coronary Angiography*) allows for increase in a reliability of state assessment of vessel wall and profile of cross section [14]. Increase in credibility of coronary angiography (CAG) can also be obtained by suitable use of *Computed Tomographic Angiography* (CTA) [1].

Main disadvantage of the method is acquiring different images depending on angle of angiographic projection, what is the cause of uncertainty in interpretation of imaging, in particular in case of presence of extensive stenoses of cardiac arteries, where possible contours of cross section through given stenosis may exhibit severe eccentricity (usually they are assumed to be irregular, in fact) as well as heavy reduction of the lumen of the vessel.

Unfortunately, above mentioned methods of visualisation of coronary arteries presents approximate only path and cross section of the vessel, involving some level of fuzziness of resulting assessment. Thus additional visualisation techniques ought to be developed and employed in order to enable cardiologists to evaluate and visualise possible deviations from calculated vessel surface and thickness of its wall and atheromatous plaques. Eventually this leads to estimation of uncertainty (upper and lower boundary) of FFR value (Fractional Flow Reserve [1]).

8.2.3 DICOM Format Versus STL

In order to satisfy the need of recording of medical imaging data dedicated standards has been established allowing for interoperability between applications and systems as well as enabling or facilitate data interpretation and visualisation also for telemedicine.[10]

In the first place DICOM (*Digital Imaging and Communications in Medicine*) standard should be mentioned, which has been elaborated in 1983 commonly by *American College of Radiology* (ACR) and *National Electrical Manufacturers Association* (NEMA) organisations [15]).[11] Popularity of DICOM (in almost all branches of medicine) emerges from its simplicity and conformance with

[10]Ex. cardiac, stomatologic, radiologic, orthopaedic consultations.

[11]Published as *ACR-NEMA Standards Publication* No. 300–1985.

requirements of other standardising agencies; ao. CEN, JIRA, IEEE, HL7 and ANSI, what without doubts positively influenced coherency and integrity of the standard.

DICOM standard ensures high quality visualisation of medical data both static and dynamic. Besides imaging data, supplementary data are recorded in relevant attributes (exemplary parameters ranges), what enables extensive postprocessing and unique identification of the patient. Simultaneously *open*[12] structure of STL format allows for processing data by third party applications, systems or devices.

These remarks are also valid for integration of information produced by dedicated applications of diverse kinds (various hardware/software environments), delivering data for processing in heterogeneous systems dedicated for storage and processing of medical data (EHR systems, *Electronic Health Record*). DICOM specification defines also requirements for data transfer, storage and accessibility by various EHR systems.

Despite many advantages,[13] visible ao. during transfer of patient's data between various medical centres equipped with imaging systems delivered by diverse manufacturers, and, what is related, diverse applications for processing and data archiving, as a main drawback of the standard, possibility of supplying redundant or useless data is mentioned. At the same time definitions of graphical objects may remain incomplete, due to filling improper fields or supplying non relevant or erroneous data or even omitting crucial fields. Another bunch of problems is connected with data processing by apparatus calibrated with different sets of parameters (ex. amplitude ranges) what deteriorates quality of images, for example by affecting of contrast or gamma profiles [16].

Nowadays, read-out of images recorded using DICOM standard is possible not only by apparatus (mainly medical) used to produce them, but also, due to openness of the specification, by third party applications[14] or by general use packages employing suitable plugins.[15] However many popular applications which are able to deal with DICOM files, converts them on import to their own working format, what can lead to difficulties during analysis including loss of information or quality.[16]

It has to be explicitly stressed that DICOM files stores spatial information mainly in raster format, it is, they comprise several, and often numerous, 2-dimensional grayscale raster images (*layers, slices*) produced by a variety of imaging systems like MRI, CT, CAT, CTA, CAG and supplementary information on their mutual layout. By the rule, these layers are parallel and laid one over other, separated by constant distance resulting from scanning resolution used (see Fig. 8.13). This is why raw DICOM files, ensuring important diagnostic capabilities, simultaneously are useless

[12] As opposed to *closed* usually *proprietary* specifications.

[13] Versatility i.e. support for all medical disciplines, high quality of data, interoperability.

[14] Ex. MatLab (functions: *dicomread*, *dicomwrite*), Photoshop Extended.

[15] Plugin is a small program or library dynamically linked on runtime extending functionality of main application (host).

[16] A good example is GIMP graphics editor which can import images from DICOM files but at the same time drops crucial data on their geometry and topology.

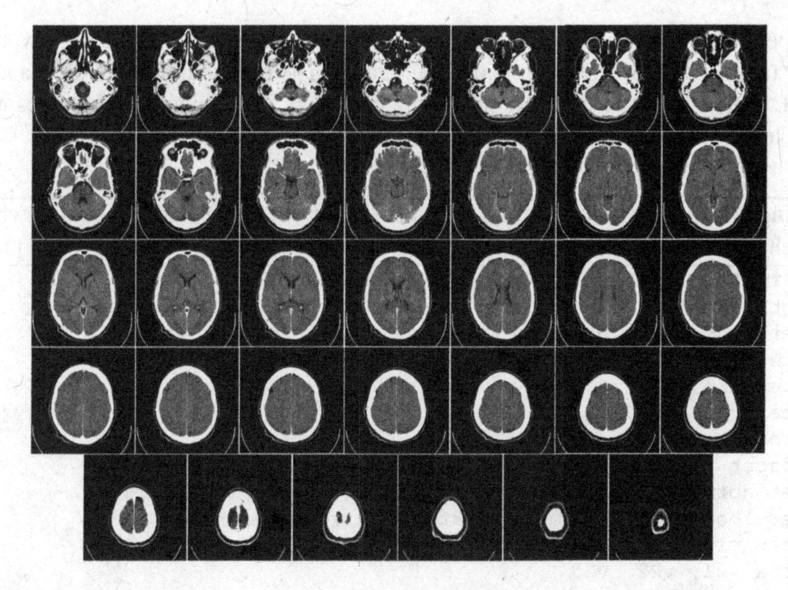

Fig. 8.13 Subsequent layers of scan of a human head (*Source* [17])

for automated, computer analysis of spatial relations, especially for assessment of state of coronary artery.

Initially STL (*STereoLithography*) format [18] has been developed years ago as input format for 3D printing systems [19].[17] Presently, this format is supported by many software packages,[18] proprietary as well as Free Software (FLOSS—*Free Libre/Open Source Software*) and is used extensively ao. for cardiology diagnostics.

STL standard can be classified as a 3D vector graphics format. The surface of the object is defined in a 3D space as plain set of triangles (tessellation). The STL definition comprises both textual (ASCII) as well as binary variant [20]. The surface is composed from triangular *facets*[19] coded separately (thus coding is redundant). Definition of every single facet includes vector normal to the surface of the triangle, directed "outside" of the object, and three additional vectors defining three vertices of the triangle ordered counterclockwise, when watching in the direction opposite to the normal vector.[20]

---

[17]For example implants and prosthetic devices or its parts can be manufactured this way.

[18]Ex. Blender, application MeshLab (http://meshlab.sourceforge.net/ revealed 16.11.2013) and many other including CAD related software (*Computer Aided Design*).

[19]This term is adopted from jewellery and means one of the several flat polished surfaces cut on a gemstone (exemplary diamond, brilliant cut), or occurring naturally on a crystal. Facets usually are flat polygons. In computer graphics all surfaces, even very complex, are built from of triangle facets (sometimes counted in myriads) in order to minimise rendering times.

[20]As an vector graphics format, STL allows for detailed visualisation of focused parts of an analysed object.

Specification of ASCII-STL file is presented in Listings 8.1 and 8.2. Key words like "**facet**" have to be written with small letters. The first item in STL file is a key word "**solid**" followed by name of elaborated object. This name may, but does not have to be used later. The last item in the file is key word "**endsolid**".

Listing 8.1 Specification of ASCII-STL file format, the text including and following "#" character comprises author's comment and is not part of STL file. Source: [18]

```
solid Description # file header
facet normal n1, n2, n3 # declaration of the first facet
outer loop # begin of declaration of vertices of facet
vertex v11, v12, v13
vertex v21, v22, v23
vertex v31, v32, v33
endloop # end of declaration of vertices
endfacet # end of declaration of facet
facet normal n1, n2, n3 # declaration of the second facet
outer loop # begin of declaration of vertices of facet
vertex v11, v12, v13
vertex v21, v22, v23
vertex v31, v32, v33
endloop # end of declaration of vertices
endfacet # end of declaration of facet
facet normal n1, n2, n3 # declaration of the next facet
outer loop # begin of declaration of vertices of facet
vertex v11, v12, v13
vertex v21, v22, v23
vertex v31, v32, v33
endloop # end of declaration of vertices
endfacet # end of declaration of facet
# all remaining facets follows
endsolid # End of file
```

Next keyword—**facet**—begins declaration of the first triangle which is finished by keyword **endfacet**. Number of facets in the STL file may easily reach and even exceed hundreds of thousands. **facet** keyword is followed by components of the normal vector. Then there is a definition of three vertices in the section beginning with words **outer loop** and terminated by words **end loop**. Declaration of components of every given vertex is open by keyword **vertex**, then follows three floats, and end of line character (**EOL** in short, in fact it is represented on Windows systems by two characters, CR and LF[21]) is expected. While original definition of the standard forbids the use of negative values of vertices components, it is not always obeyed. It is worth to say, that this restriction is artificial and is not derived from any programming principle.

[21]CR-carriage return (ASCII '13'), LF—line feed (ASCII '10'). On Unix systems there is always one character, namely **CR**.

Listing 8.2 Example of ASCII-STL file. Words formatted in italics can be edited.

```
solid ASCIIFile
facet normal 0.940 −0.039 0.337
outer loop
vertex −38.000 −18.549 −8.320
vertex −37.970 −18.496 −8.399
vertex −37.990 −18.367 −8.326
endloop
endfacet
facet normal 0.941 −0.040 0.333
outer loop
vertex −38.000 −18.549 −8.320
vertex −37.990 −18.367 −8.326
vertex −38.020 −18.420 −8.247
endloop
endfacet
facet normal 0.921 0.0503 0.385
outer loop
vertex −38.02 −18.42 −8.24
vertex −37.99 −18.36 −8.32
vertex −38.02 −18.23 −8.25
endloop
endfacet
...
...
endsolid
```

In "**outer loop**" section at least three vertices have to be present. Though STL format specification allows for more than three vertices, usually only three (i.e. triangles) are used, what enhances efficiency of calculations. Subsequent numbers (floats with decimal **points**) are separated by spaces (at least one space, ASCII code '32'), keywords are put without quotations around. There is no scale definition in the file and units are relative. Program performing tessellation (generating facets) has to define them in such a way that vertices of one facet are not on the edge of another. So, if there are two triangle neighbours, they have exactly one common vertex or one whole edge. Due to round up errors a couple of vertices belonging to the neighbouring triangles may not coincide, thus a tiny crack may be spotted in the tessellated surface.

As the above description suggests, structure of STL file resembles XML file structure, however tags in STL file are not enclosed in acute parentheses. It should be noted, that STL standard had been elaborated long ago XML draft emerged.

As stated in the previous paragraph, STL format defines object geometry only and is relatively ascetic—contains three even tags (a pair of opening and closing tag) and one odd tag (opening only). Simplicity of the definition is clearly advantage of the STL format. Further, this format is suitable for processing on contemporary, fast graphics cards, designed for rendering hundreds of thousands of triangles per second.

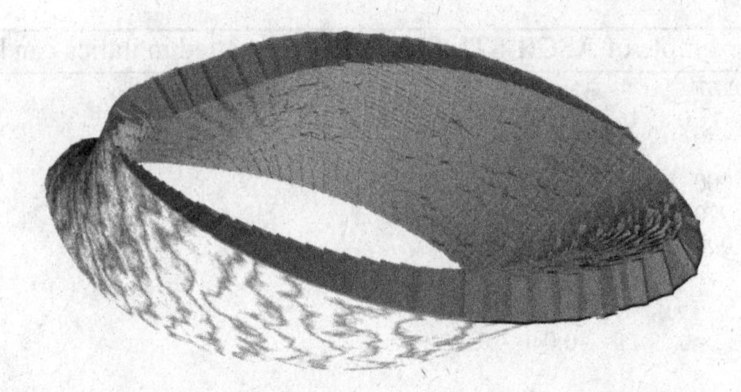

Fig. 8.14 Element of cardiac vessel rendered using "thick facets". On the *internal surface* visible is pattern created by individual prisms. *External surface* is patterned using marble like texture and by no means reflects real look of blood arteries. *Source* visualisation produced using STL.inc module from ScPovPlot3D package

In particular after few adjustments can be supplied as input data file to thick surface procedures, implemented in STL.inc library from mentioned above ScPovPlot3D package. Example of resulting image of cut-out of coronary vessel is given in the Fig. 8.14.

Despite that STL format has severe shortcomings, it is satisfactory as a base for visualisation of 3D fuzzy surface. However, coronary vessel can be visualised only in tiny slices in order to avoid screening of one part by another. Whole vessel may be then presented using animation created in automated or interactive manner. The latter requires additional presentation software, which is beyond of the scope of this book.

References

1. Taylor, C.A., T.A. Fonte, and J.K. Min. 2013. Computational fluid dynamics applied to cardiac computed tomography for noninvasive quantification of fractional flow reserve: Scientific basis. *Journal of the American College of Cardiology* 61(22): 2233–2241.
2. Opiła J., I. Skalna, and T. Pełech-Pilichowski. Interval arithmetic for irregularly distributed data visualisation by Kriging. In *XIV International scientific conference*, editor, *Corporate Governance—Theory and Practice*, 1–14, Krakow, 22–23 Nov 2012. WZ AGH.
3. StockCharts.com, Inc. Japanese candlesticks, introduction to candlesticks. http://stockcharts.com/school/doku.php?id=chart_school:chart_analysis:introduction_to_candlesticks. Accessed 25 Sept 2014.
4. Spruyt, V. How to draw a covariance error ellipse? http://www.visiondummy.com/2014/04/draw-error-ellipse-representing-covariance-matrix/. Accessed 29 Oct 2014.
5. Jenness, J., and L. Engelman. Jennes enterprises, ArcView. http://www.jennessent.com/arcview/mahalanobis_description.htm. Accessed 29 Oct 2014.
6. Mahalanobis, P.C. 1936. On the generalised distance in statistics. In *Proceedings of the National Institute of Sciences of India* 2: 49–55.
7. Krige, D.G. 1951. A statistical approach to some basic mine valuation problems on the witwatersrand. *Journal of the Chemical, Metallurgical and Mining Society* 52: 119–139.

8. Art and animation studio 2015, GPU vs CPU. GPU rendering vs software CPU rendering. http://furryball.aaa-studio.eu/aboutFurryBall/compare.html. Accessed 25 Nov 2014.
9. Achenbach, S., J. Walecki, M. Zawadzki, and M. Witulski. 2006. Clinical applications of multislice computed tomography in cardiology. *Postępy w Kardiologii Interwencyjnej* 2(2): 160–168.
10. Legutko, J., J. Jąkała, B. Mrevlje, S. Bartuś, and D. Dudek. 2011. Fractional flow reserve-guided myocardial revascularization. *Postępy w Kardiologii Interwencyjnej* 7(3 (25)): 228–241.
11. Bartoszek, B., M. Mościński, T. Niklewski, B. Szyguła-Jurkiewicz, and A. Lekston. 2011. Wirtualna histologia—nowoczesna metoda oceny tętnic wieńcowych. *Folia Cardiologica Excerpta* 6(3): 203–209.
12. Kubica, J., R. Gil, and P. Pieniążek. 2005. Wytyczne dotyczące koronarografii. *Kardiologia Polska* 63(5(3)): 491–500.
13. Paluszek, P., P. Pieniążek, P. Musiałek, T. Przewłocki, A. Kabłak-Ziembicka, R. Motyl, Ł. Tekieli, A. Leśniak-Sobelga, K. Żmudka, and W. Tracz. 2009. Symptomatic vertebral artery stenting with the use of bare metal and drug eluting stents. *Kardiologia Polska* 5(1(15)): 1–6.
14. Barańska-Kosakowska, A., M. Hawranek, M. Gąsior, A. Spatuszko, R. Przybylski, M. Zakliczyński, and M. Zembala. 2010. Serial quantitative coronary angiography (QCA) in the assessment of transplant coronary artery disease (TxCAD). *Kardiochirurgia i Torakochirurgia Polska* 7(3): 319–324.
15. Digital imaging and communications in medicine. Strategic document (last revised 23.10.2013). http://medical.nema.org/dicom/geninfo/Strategy.pdf. Accessed 14 Nov 2013.
16. Mustra, M., K. Delac, and M. Grgic. 2008. Overview of the DICOM standard. *50th international symposium ELMAR-2008*, vol. 1, 39–44. Croatia.
17. Haeggstroem, M. 2008. Computed tomography of human brain. *Uppsala University Hospital*. http://en.wikipedia.org/wiki/File:Computed_tomography_of_human_brain_-_large.png. Accessed 10 May 2015.
18. Burns, M. The STL format. Standard data format for fabbers. reprinted from section 6.5 of automated fabrication, Ennex corporation. http://www.ennex.com/~fabbers/StL.asp. Accessed 14 Nov 2013.
19. Palermo, E. What is stereolithography? Live science. http://www.livescience.com/38190-stereolithography.html. Accessed 16 Nov 2013.
20. Burns, M. 1993. *Automated fabrication: Improving productivity in manufacturing*. PTR Prenrice Hall.

Printed in the United States
By Bookmasters